D0321768

BRISTOL
CITY COUNCIL

AN 1972670 8

With great appreciation, I dip me lid to my Australian connection which is led by **Slim Bauer**, **Jack Foley**, **Derek Smith**, **John Holmes**, and **Speedy Solomon**.

Contents

CHAPTER THREE

Peopling the Austral Continent, 40

CHAPTER FOUR

Fruits of the Land: The Primary Industries, 65

CHAPTER FIVE

Contemporary Australia, 88

CHAPTER SIX

Australian Regions, 104

Preface

Now that the Pacific is the ocean of choice for Americans, Australia looms larger and larger in our consciousness. Not long ago, we were an Atlantic-oriented nation, but nowadays our trade and travel are increasingly focused upon the ocean to our west. Our escalating attention to "the Pacific rim" is aimed mostly at the populous Asian nations that border the western side of the ocean. But Australia and New Zealand are also Pacific rimmers and their importance as vacation destinations and trading partners continues to grow. Thus it seems to be an appropriate time to learn as much as we can about Australia's corner of the world. To that modest goal, this book is pledged.

I have had the good fortune to spend considerable time in Australia on 10 separate visits over the past three decades. Cumulatively, four and a half years of my life have been lived "Down Under." I fell in love with Oz (as it is known in the Aussie idiom) on my first visit, when I resided for a year in Adelaide; I can still savor the pleasure of apportioning my time among the Uni's Barr-Smith Library, the Hampshire pub in Grote Street, and the outdoor handball courts of Christian Brothers College in Wakefield Street. I will forever be an Australophile, although I recognize many of the country's imperfections. If this book can contribute in a small way to disseminating a broader understanding of Australia among Yanks, I will be very gratified.

Acknowledgments

My gratitude goes out to a host of Aussies, especially those noted on the Dedication page. In addition, I would like to acknowledge:

- The role of certain geographer/administrators who made it easy for me to have temporary appointments at their universities: **Fay Gale, John Holmes, John Oliver, Tom Perry, Bruce Thom**, and, particularly, **Graham Lawton** and the late **Gilbert Butland**, who threw wide open the doors at Adelaide and at New England.
- The thoughtful critiques of the entire manuscript by **Slim Bauer** of Canberra, **Robin Kearns** of Auckland, **John Pigram** of New England, and **Joe Powell** (for the substance, not the style) of Monash.
- The good sportsmanship of **Marylee** (1929–1986), who went uncomplainingly with me down many a dirt track.
- The inspiration I received from **Gordon Lewthwaite** over the past quarter century of shared interests in the Southwest Pacific.
- The excitement of sharing Australia with **Joani**, who loves it as much as I do.

CHAPTER ONE

The Pacific Setting

The largest body of water known to humankind extends from the Arctic beaches of the Bering Sea in the north to the icy margins of Antarctica's Ross Sea in the south. On the east and west it washes the shores of four great land masses. Only Africa and Europe among the seven recognized continents do not lie on its fringes.

This vast ocean, known since the early 1500s as "the Pacific," is the most conspicuous feature of our planet. With an extent of nearly 65,000,000 square miles (168,000,000 km^2), it is more than twice the size of the second largest ocean and encompasses a greater area than all of the world's land surfaces combined. It occupies approximately one-third of the total surface of the earth.

The immensity of the Pacific is an inescapable reality, and most of its expanse is unblemished by land. Although in total there are literally tens of thousands of land masses, ranging in size from continental to minute, rising above the blue waters, these islands occupy but a fraction of the Pacific Basin. Many continent-sized portions of the ocean are completely devoid of islands, without a single piece of ground breaking the sweep of the water. It is possible to travel great distances in any part of the Pacific except the southwest without coming in sight of land for weeks or even months. Indeed, the first expedition to circumnavigate the globe, that of Magellan in 1520 to 1522, sighted only two small islands while traversing the Pacific from southernmost South America to Guam in 98 days of sailing.

The Pacific is largely a tropical ocean. Although it extends almost to the Arctic Circle in the north and a few degrees beyond the Antarctic Circle in the south, its greatest girth is in equatorial regions, so most of its islands and sea lanes lie within the tropics. Its personality, while not always as placid as its name would indicate, is characterized by such low-latitude phenomena as tepid water temperatures, persistent trade winds, towering cumulonimbus clouds, exquisite sunsets, abrupt but short-lived thunderstorms, devastating tropical cyclones, extensive coral reefs that separate pounding surf from peaceful lagoons, brilliant beaches inevitably bordered by towering palms, and greatly varied marine life.

The Oceanic Influence

The sea, then, is the one ineluctable fact of this region. Its influence is dominant, and only in the interior of large land masses (Australia, New Guinea, New Zealand) is it possible to escape its pervasiveness. This influence is felt in many ways.

The ocean is a barrier in the basic sense. It has prevented the spread of most types of terrestrial life. Only a relatively few varieties of plants and animals can survive a long trip from one land area to another:

- Those with seeds that are watertight and buoyant (e.g., pandanus);
- Those that can live comfortably in driftwood (e.g., weevils and lizards);
- Those light enough to be wafted long distances by the wind (e.g., fern spores or small spiders); or
- Those that may be carried inadvertently by flying creatures (e.g., mites on bats or small arthropods on birds).

Thus, the possibilities for natural diffusion across this vast ocean are limited, resulting in the broadly valid generalization that the further one goes east and north from Southeast Asia, the less diverse and rich the biota becomes.

On the other hand, the ocean serves as a link between lands that are thousands of miles apart. The buoyancy of water enables relatively heavy objects to float for indefinite periods, and the persistence of both ocean and air currents provides a means of locomotion for a floating object. Of all the species of the planet, this connectivity function is most significant for *Homo sapiens*. Even with relatively frail crafts, humans have sailed to the remotest reaches of the oceans. And where they have gone, they have taken with them plants and animals, both cultigens and parasites, affording further means for diffusion. Thus people and their camp followers (such as dogs, pigs, rats, fleas, sweet potatoes, taroes, etc.) have dispersed over the islands of the Pacific, investigating them all and settling on all but the poorest.

Wherever humans have settled in the Pacific, the ocean has continued to exercise a prominent influence on life. It contributes to the economy in both subsistence and commercial trade by providing a habitat for food resources. Although tropical waters generally cannot match the great quantity of fish found in certain mid-latitude oceans, the variety of marine life around Pacific islands is unparalleled. On most Pacific islands, coastal fishing occupies a significant niche in the way of life.

Inter-island communication provides another focus for assessing the influence of the ocean. Trading and raiding; colonization and warfare; exploitation and missionizing; the sea provides an avenue for these and other human interactions. In the past, Tongans sailed to Fiji to learn the art of warfare, ni-Vanuatu were "blackbirded" to work in Australian sugar cane fields, Pitcairners were moved to the more permissive environment of Norfolk Island, Hawaii was "unified" by Kamehameha's flotilla of war canoes advancing northwestward from the Big Island. Today, Cook Islanders migrate to New Zealand, Nauruans export phosphate to the world, Japanese tuna boats support the Samoan economy, and overcrowded Tuvaluans are resettled in Fiji.

Whereas its roles as barrier, as link, as source of food, and as communication medium are all of major importance, it is likely that the influence of the Pacific Ocean on climate is most significant of all. The physical nature of the Pacific Basin is in large measure determined by the characteristics of the water and their concomitant effects on the overlying air. The southern part of Australia and most of New Zealand are dominated by midlatitude westerly winds, but no other significantly inhabited Pacific area is influenced by cool air masses. Most of the Pacific lands are dominated by tropical airflow, particularly by the easterly trade winds. These wind systems demonstrate characteristics acquired from the warm waters over which they blow; thus warm, relatively humid conditions are to be expected over most of the region most of the time. Precipitation is variable, but in general it is abundant. There is tremendous potential for rainfall, and unstable lapse rates often develop, resulting in very heavy rains, both seasonal and nonperiodic. Thunderstorms are widespread, tropical cyclones are felt in many parts of the region, and milder tropical disturbances also occur. Spectacular clouds, thunder, lightning, and rainbows are commonplace. Even on the so-called "desert" islands, which are small coralline outcrops with so little relief and area that they do not trigger air mass uplift and showers, there is an abundance of moisture in the form of humidity. In most Pacific island areas, then, land and water and air share the uniform attributes of warmth and wetness.

Land Masses of the Pacific

Despite the vastness of its water surface and the importance of its sea lanes and fisheries, it is not the ocean itself that draws most of our attention; rather, we focus upon the relatively small amount of land that is scattered over the Pacific, for this is where people live in varying densities and patterns. In keeping with this book's concentration on Australasia, the continental fringe islands around the margin of the Pacific will be excluded. The near-coastal islands of the Americas (the Chilean islands, the Galapagos, the offshore Mexican islands, California's Channel Islands, the coastal islands of British Columbia and Alaska, and the Aleutians) are more properly considered in conjunction with the countries they border; and the significant islands

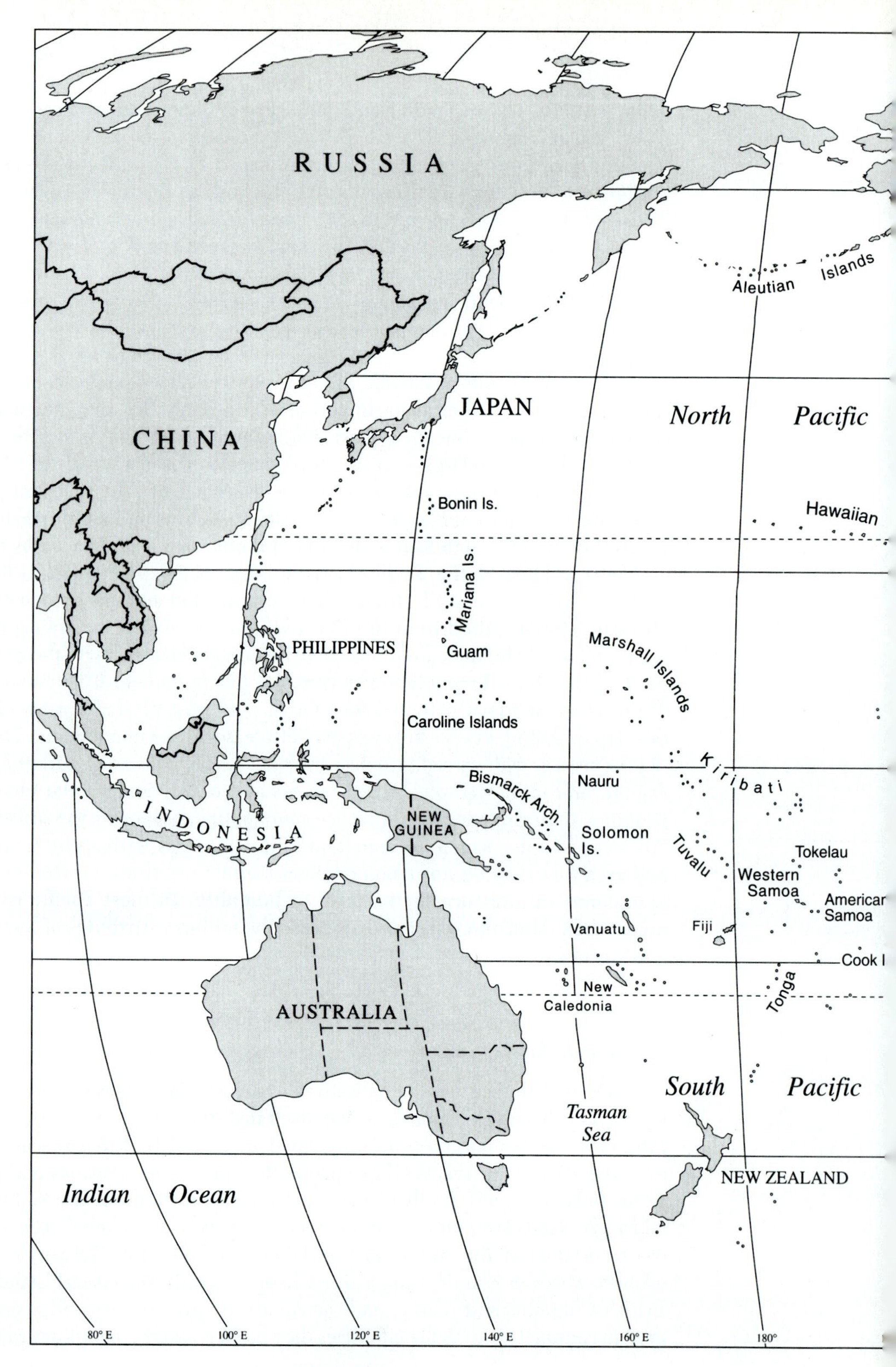

Figure 1-1 The Pacific Setting

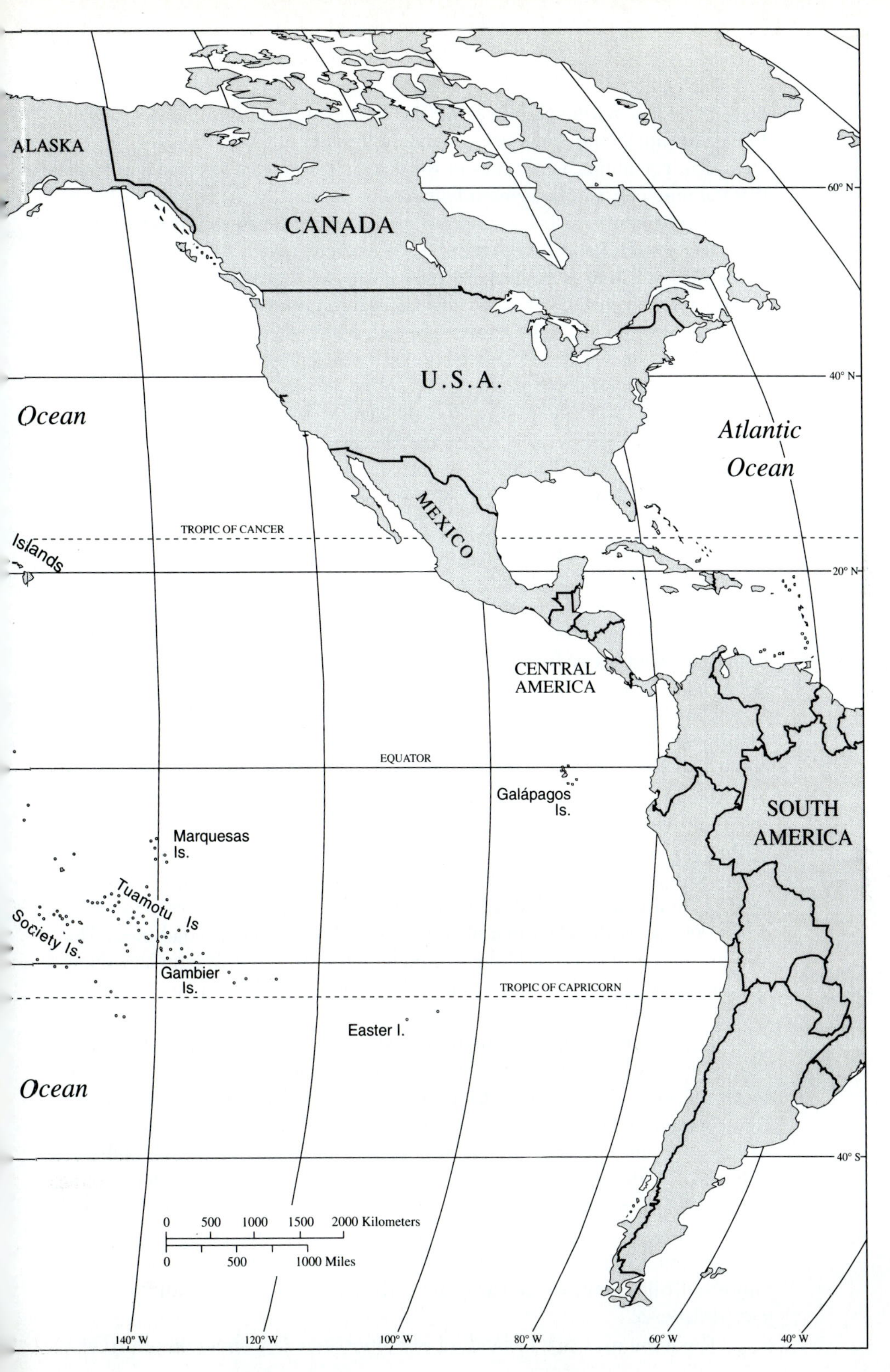
ALASKA
CANADA
U.S.A.
Ocean
Atlantic
Ocean
MEXICO
TROPIC OF CANCER
Islands
CENTRAL
AMERICA
EQUATOR
Galápagos
Is.
SOUTH
AMERICA
Marquesas
Is.
Tuamotu Is
Society Is.
Gambier
Is.
TROPIC OF CAPRICORN
Easter I.
Ocean
60° N
40° N
20° N
40° S
0 500 1000 1500 2000 Kilometers
0 500 1000 Miles
140° W
120° W
100° W
80° W
60° W
40° W

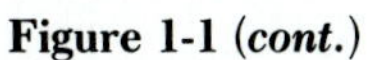

Figure 1-1 (*cont.*)

on the western margin of the Pacific (the Kuriles, Japan, the Ryukus, Taiwan, the Philippines, and Indonesia) are integral parts of Asia.

The remaining land areas of the Pacific (those that are the subject of this volume) consist of a tremendous number of mostly tiny tropical islands, a small continent (Australia) that has a subtropical location but whose low-latitude characteristics are sometimes obscured, and a large island group (New Zealand) that is subtropical only on its northern margin. These land masses are distributed very unevenly throughout the Pacific, with the principal concentration in the southwest portion. Most of the northern part of the ocean, between California and Japan east-west and between the Aleutians and Midway Island north-south, is absolutely devoid of any land rising above the surface of the sea. An equally extensive expanse of the eastern Pacific, stretching from the west coast of the Western Hemisphere continents to about the 135th meridian west, is dotted with less than a dozen tiny isles. In its south-central and southwestern portions, however, the ocean is peppered with islands.

This multitude of islands lends itself to a convenient, and traditional, threefold subdivision arising from the etymological roots of the island groups. The "black islands" of Melanesia are all in the southwest Pacific; the "many islands" of Polynesia occupy a vast expanse of the central Pacific; the "small islands" of Micronesia lie mostly north of the equator and west of the International Date Line.

The Melanesian realm of the southwestern Pacific is separated from Micronesia to the north by an extensive equatorial expanse of island-free ocean and from Polynesia to the east by a similar spread of landless water. The Melanesian core is characterized by "continental" islands, whose rugged relief and volcanic origins impart a distinctive character to the land masses of this part of the Pacific basin. Most notable is New Guinea, whose area (286,000 square miles [462,000 km^2] in the Melanesian portion) is exceeded by that of only one other island in the world, and whose population (nearing 4,000,000) is greater than that of all the other Pacific islands combined. The other important islands of Melanesia include the Bismarck archipelago (New Britain, New Ireland, and the Admiralty Islands), the Solomons, Vanuatu, and New Caledonia.

The great triangle of Polynesia, with its corners at Hawaii, Easter Island, and New Zealand, sprawls across 60° of longitude (mostly east of the 180th meridian) and 50° of latitude (mostly south of the equator). Excluding New Zealand (which will be dealt with separately) and Hawaii (which is not within the purview of this volume), about half a million people inhabit Polynesia. Although the vast majority of the islands are small, flat atolls, some are high, rugged, and volcanic. The principal island groups in Polynesia include the Line, Phoenix, Ellice, Tokelau, Samoan, Tongan, Cook, Society, Tuamotu, and Marquesas islands. At the western edge of Polynesia lies the important Fiji group, which has many of the characteristics of both Polynesia and Melanesia.

The numerous small islands of Micronesia contain about one-third of a million inhabitants. A complicated political structure fragments the four principal groups of islands: the Marianas, the Carolines, the Marshalls, and the Gilberts. Nearly all of the individual islands consist of coralline atolls or outcrops of quite small size.

Located far to the south of the rest of Polynesia, and well beyond the margins of the true tropics, is New Zealand. This southernmost nation in the Eastern Hemisphere consists of two principal islands (named the South and the North), which are the twelfth and fourteenth largest in the world, as well as much smaller Stewart Island in the far south. New Zealand's high relief and generally rugged terrain reflect its location at the confluence of two major submarine mountain arcs, one extending southeastward from New Guinea and New Caledonia and the other southward from Fiji.

The largest land mass of the Pacific area is the continent of Australia, which is approximately the same size as the 48 conterminous states of the United States. Situated 1,400 miles (2,240 km) northeast of New Zealand, 1,000 miles (1,600 km) southwest of New Caledonia, and less than one hundred miles (160 km) south of New Guinea, Australia is characterized by relatively level terrain and dry climate. Thus its environmental characteristics differ significantly from those of the other land areas of the Pacific Basin.

CHAPTER TWO

Australia: The Unique Land

Australia is the smallest of the world's seven continents and the only one that is comprised of a single political unit. With an area of 2,971,081 square miles (4,753,730 km^2), it ranks as the sixth largest country in the world. It occupies an essentially transitional position between the low and middle latitudes, almost 40% of its area being within the tropics. Thus its physical characteristics are reflective of a large, relatively compact land mass in the subtropics, although the great extent of aridity and the concentration of population in the least tropical portions of the country tend to deemphasize this subtropicality. Australia's greatest poleward extent is the southern end of the island of Tasmania, which reaches beyond 43° south latitude, roughly as far from the equator as Boston, Buffalo, or Detroit.

Whereas there is nothing remarkable about the geographical position or dimensions of Australia, most other aspects of its physical geography have unusual characteristics that tend to set it apart from the other continents in a variety of ways. The other five inhabited continents exhibit a certain regularity in patterns of physical geography, which give them an element of uniformity and predictability. Generally, position on the continent (encompassing both latitude and nearness to the coast) and various broad distribution patterns of climate, soils, natural vegetation, and native animal life are associated. The persistence of these environmental patterns from continent to continent is a notable feature of the physical geography of the world. In Australia, however, the patterns are interrupted, rearranged, and otherwise modified.

The Tectonic Background

It was perhaps a billion years ago that ancient continental surfaces (called *cratons*) were welded together by massive collision to form the western two-thirds of the Australian landmass as we know it today. This was part of the Gondwana supercontinent that included the mass of India/Tibet to the west and Antarctica to the south.

Over eons of time the relatively rigid surface crustal plates slowly broke apart and drifted independently over the underlying hot, plastic asthenosphere layer. The Australian tectonic plate separated from the Indian and Antarctic plates and sluggishly drifted northeasterly, where it collided with the southwestward-moving Pacific plate. The latter plunged beneath Australia, resulting in the addition of new land to Australia's eastern side. By about 40 million years ago marine barriers were wide and well developed around Australia, and it had indeed become the "island continent."

More recently (ca. 25 million years ago), the Australian plate's northward drift impacted it with an Asian island chain. This collision produced the mountainous backbone of New Guinea, while downwarping south of the collision zone left a large area—the Arafura Sea, Torres Strait, and the Gulf of Carpentaria—submerged just north of the Australian continent.

The Shape of the Land

The general shape of continental Australia can be likened to an east-west oriented football that has been severely constricted on its mid-southern and northeastern margins. The Great Australian Bight comprises an enormous, but relatively shallow, embayment in the southern coast, and the Gulf of Carpentaria makes a smaller but sharper recess in the northeast (see Figure 2-1). Apart from these two major indentations, the coastline is fairly smooth and regular in gross outline, although there are many minor irregularities. Islands occur in varying profusion around the margins of the continent, but most of them are quite small. Most important by far is the island state of Tasmania in the southeast, which is twenty-fifth in size among the islands of the world. Other significant islands include King Island and the Furneaux Group in Bass Strait between Tasmania and the mainland, Kangaroo Island off the coast of South Australia, Melville and Bathurst islands off the mid-north coast, several groups in the Gulf of Carpentaria (Groote Eylandt, the Sir Edward Pellew Group, and the Wellesley Islands), the Torres Strait Islands off the northern tip of Queensland, and the varied islands and islets associated with the Great Barrier Reef.

Probably the most important basic generalization about the Australian lithosphere is that the continent is an ancient and stable one. Throughout eons of geologic time it has demonstrated the sort of long-range stability that characterizes the better-known "shield" areas of the Northern Hemisphere (Laurentian, Fenno-scandian, Angara). That is to say, for the most part, major crustal movements have been rare and vulcanism has been restricted to a few regions. Much of the present Australian continent, then, has been in a quiescent, rigid, subaerial situation for millions of years, and this condition has had a marked effect upon the landscape.

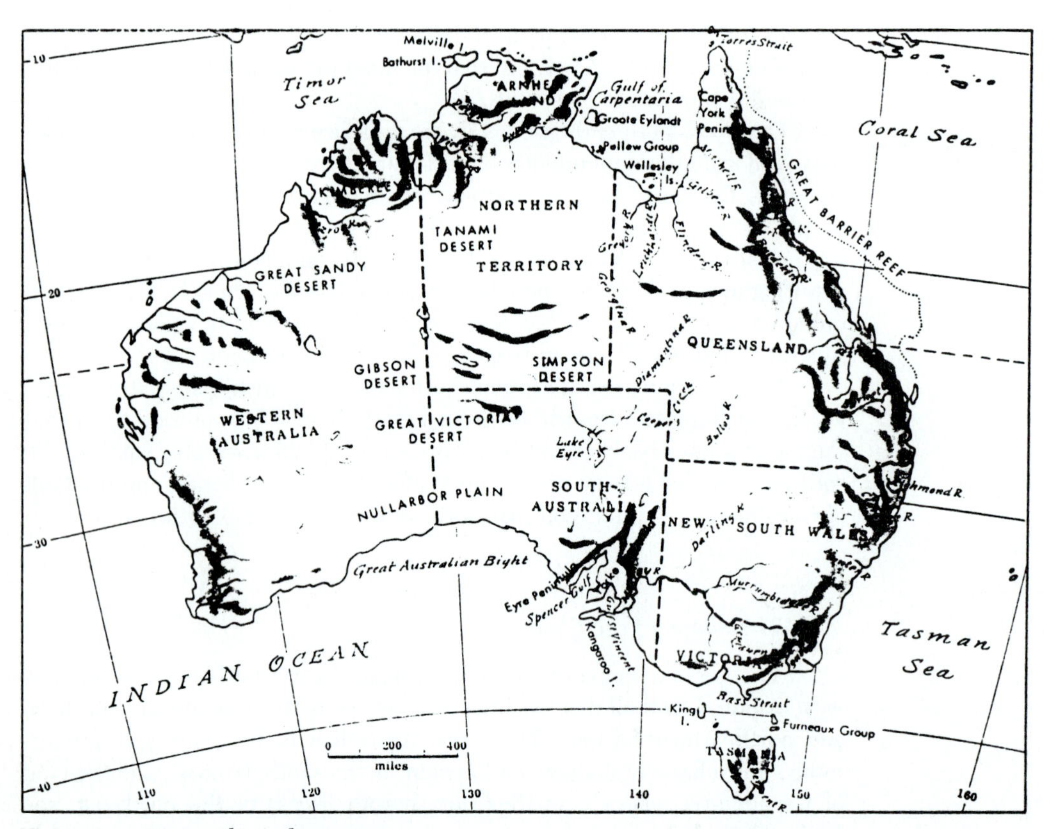

Figure 2-1 Major Physical Features

In a very real sense Australia is one of the oldest of the continents. It has been subjected to lengthy, relatively uninterrupted periods of weathering and erosion that have subdued its highlands and reduced its surface to one of low elevation and relief. Spectacular topographic features are rare. Only around the margins, particularly in the east, and in a few scattered localities in the interior has there been significant recent orogenic activity resulting in notable mountains. For the most part Australia is replete with desert landforms, most of which are the result of long-continued differential erosion, rather than orogeny. Such features as massive knobs and monoliths and domes with exfoliated surfaces, caprock escarpments, intermittent stream beds, basins of interior drainage with playas and salinas, vast linear dunefields, and gibber (angular rock fragments) plains are characteristic.

The Eastern Highlands. Extending the entire length of extreme eastern Australia, from northern Queensland to southern Tasmania, is the subdued cordillera of the Eastern Highlands, often referred to as the Great Dividing Range (see Figure 2-2a). The highlands had a moderately complex origin, involving a mixture of folding, faulting, warping, igneous intrusions, and

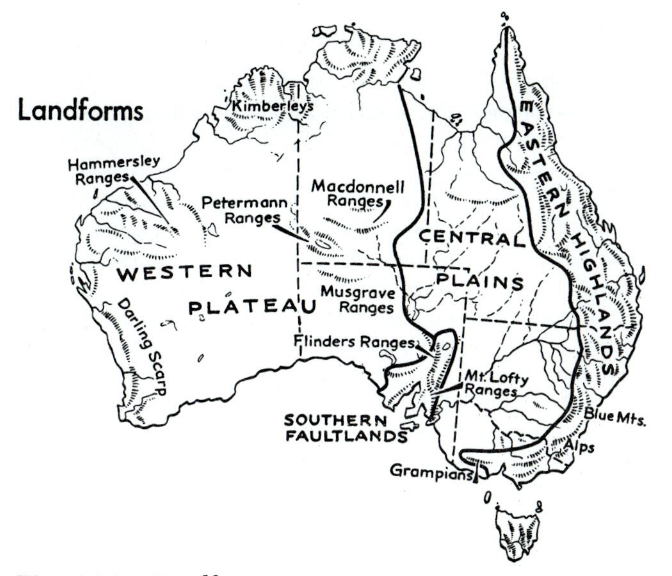

Figure 2-2a Landforms

limited volcanic extrusions, with the majority of the mountain-building activity occurring in late Tertiary time. Most of the highest peaks consist of intruded and uplifted granite masses.

The topographic variety encompassed in the cordillera is easily the greatest in Australia. In many places, particularly toward the north, the mountains are so low and subdued that they are quite inconspicuous. In other localities there are massive escarpments, deep gorges, and sheer cliffs that almost defy penetration. On the eastern margin many of the highland ridges project to the sea, descending steeply to the water's edge; however, in most areas there is a narrow fringe of flattish coastland made up of a series of small river valleys, but nowhere of large enough size to be thought of as a coastal plain.

In Queensland the highlands are mostly of low or moderate height, the crests averaging between 2,000 and 3,000 feet (600 and 900 m) above sea level; but in some places the highland area is quite broad. There are some conspicuous peaks and scarps, but mostly low hills are typical. On the eastern side the highlands are interlaced with relatively small, flattish riverine plains, whereas on the west there is a gradual merger with the lowlands of the interior (see Photo 2-1).

In New South Wales the highland is somewhat more restricted in area, but there is a general increase in elevation, relief, and diversity of landforms. The Northern Tablelands are separated from the coast by an irregular but remarkably steep and abrupt escarpment, which is one of the most spectacular terrain features in Australia. In the central portion of the New South Wales highlands is the maze of the Blue Mountains, whose characteristically flat valleys and ridge tops are separated by steeply plunging slopes. In

Focus Box: The Great Barrier Reef

Off the northeastern coast of the continent is the largest and longest coral formation in the world, called the Great Barrier Reef. Despite its name, this is not a single extensive reef; rather it is a complex assemblage of more than 2,500 individual reefs that vary in size from isolated pinnacles to massive structures up to 15 miles (25 km) long and 50 square miles (130 km^2) in area. Some 320 coral islands (nearly all of them quite small) are interspersed among the reefs. This enormous "reef province" (as scientists prefer to call it) trends generally north-south for 1,200 miles (2,000 km), covering about 15° of latitude and extending over some 90,000 square miles (230,000 km^2) of ocean. At some points it is only a few miles from the mainland; at others it is almost 200 miles (320 km) offshore.

A coral reef is formed when tiny anthozoan animals called coral polyps attach themselves to the shallow sea bottom and secrete limy external skeletons around the lower half of their bodies. These tiny creatures (most are only a fraction of an inch long) live in colonies of countless individuals, attaching themselves to one another both with living tissue and with their external skeletons. Their extraordinary abundance is a tribute to their remarkable reproductive capabilities, because they are not actually very hardy creatures. They cannot survive in water that is very cool or very fresh or very dirty. Moreover, they require considerable light, so they cannot live more than a few dozen feet below the surface of the ocean.

Under the best of circumstances, a coral colony cannot build a reef unaided. The many openings in and among the skeletons must be filled in and cemented together by other agencies. Much of the infilling material consists of fine sediment that is provided by other marine animals through their borings and waste products, and a host of tiny animals and plants contribute their skeletons and shells when they die. Indispensable to the reef-forming process are cement and mortar provided by algae and other tiny forms of "seaweed."

As with most reefs, the Great Barrier Reef was formed on a slowly sinking seabed. As sinking progressed, the deeper corals died, their skeletons providing a firm foundation upon which more polyps secreted more skeletons in the shallower "live" portion of the reef. Near the surface, the reef-building flattens and expands horizontally. The resultant "barrier" interrupts wave action and water circulation, creating different environments on each side of the reef. The protected inner (shoreward) side is bathed in warmer, placid waters, whereas the outer side receives the full force of the pounding sea.

Although corals are the dominant creatures of the reef, a great variety of other animals (as well as plants) share the habitat, making a coral reef one of the richest ecosystems known. Mollusks, such as clams, cowries, cones, snails, sea slugs, squids, and octopuses, are particularly common. Also notable are such crustaceans as shrimp, crabs, lobsters, crayfish, and echinoderms (such as starfish and sand dollars). Fishes occupy the reef environment in enormous numbers and variety. Other common vertebrates include sea turtles, sea snakes, dolphins, and whales. And the small coral islands are home to vast numbers of sea birds.

Most of the Great Barrier Reef has been set aside by the Australian government as a national marine park, and is recognized by the International Union for the Conservation of Nature as a World Heritage Site. The reef is one of Australia's most compelling tourist attractions, for both domestic and international visitors. Some 30 resorts have been established on reef islands. More than 200,000 people stayed overnight at one of the reef resorts in 1990, and many others made day trips to the reef.

southernmost New South Wales the highland mass reaches its greatest heights, with the Australian Alps extending into Victoria. True alpine terrain is lacking, but an uplifted and eroded tableland presents certain spectacular aspects, and there is clear evidence of past glaciation.

In Victoria the highlands trend to the west with considerable grandeur, encompassing some of the wildest and least accessible terrain on the continent. In the central part of the state the ranges become more subdued, often little more than dissected cuestas, finally terminating near the South Australian border in a discrete highland called the Grampians.

Tasmania contains much topographic variety but is geologically related to, and should be considered part of, the Eastern Highlands. In general, the island represents an uplifted dome that has been dissected and glaciated in

Photo 2-1 *The topography of the east coast of Australia mostly consists of a mixture of steep forested headlands and expansive beaches. This scene is in north Queensland, near Port Douglas. (TLM photo.)*

its higher parts. Although its relief is restricted, Tasmania is comprised of a complexity of hills and mountains, with limited areas of flat land.

The Central Plains. West of the highlands there is a gradual transition from hills to plain, as well as a continuing decline in elevation; thus the Eastern Highlands merge more or less imperceptibly with the Central Plains. This region is comprised of an extensive area of lowland, roughly extending north-south from the Gulf of Carpentaria to the Great Bight, and displaced eastward from the center of the continent (see Photo 2-2).

The plains are underlain generally by horizontal sedimentary rocks, with a considerable veneer of more recent alluvial and aeolian deposits on the surface. The lowland is broken in various places by ranges of low, rocky hills, none of which are very extensive in area. Loose deposits of sand are found in many localities, with a very expansive region of lengthy sand ridges in the southeastern corner of the Northern Territory (Simpson Desert) and smaller areas of similar development in the northeastern portion of South Australia.

Photo 2-2 *Typical flattish terrain of the Central Plains. This scene in north-central Queensland shows an abundance of termite mounds, which are commonplace in northern and central Australia. (TLM photo.)*

The Southern Faultlands. A relatively small portion of the continent, located along and near the coast of South Australia, is dominated by a parallel series of steep hills and oceanic gulfs. These are the products of complex crustal movements, primarily faulting along parallel zones of weakness, but including some complicated folding in the north. From west to east are found the varying terrain of the Eyre Peninsula, the rift valley of Spencer Gulf, the hills of Yorke Peninsula, Gulf St. Vincent, and the Mount Lofty Ranges, which merge with the higher and more rugged Flinders Ranges to the north.

The Western Plateau. Almost half of the entire continent consists of a rigid pre-Cambrian shield, surface expression of which is a low plateau. Scattered and isolated mountain ranges and higher plateau blocks interrupt the continuity of the surface, but only in three areas has there been general uplift above the plateau.

1. The highest elevations are found in the Pilbara country of the northwestern "shoulder" of Western Australia, where the Hammersley Ranges have peaks that reach to 4,000 feet (1,200 m) above sea level. This is broken sandstone country cut by several major river valleys.
2. The Kimberleys district, in the north of Western Australia, consists of a number of confused ranges, mostly composed of rugged sandstone. Although the elevations are not high (peaks up to 3,000 feet [900 m]), much of the landscape is spectacularly irregular and dissected by deep gorges.
3. The uplands of Arnhem Land, in the far north of the Northern Territory, are lower and less rugged than the two areas just discussed. Nevertheless, the terrain rises sharply above the fringing plains and is characterized by broad, incised river valleys.

There are a number of other rocky ranges in the west, but they are relatively minor features that rise like elongated islands out of the general lowland. Most notable are the Macdonnell, Musgrave, and Petermann ranges in the heart of the continent. These are composed largely of granitic and metamorphic rocks uplifted in an east-west orientation. Antecedent streams draining the ranges often cut directly through the ridges in steep, narrow gorges, which divide the ranges into relatively small blocks.

Impressive, isolated eminences rise above the general plateau level in various places. They are usually rounded granite or sandstone projections with conspicuous exfoliation. The most spectacular are found in the southwestern corner of the Northern Territory, in the form of monolithic *bornhardts* (rounded or domal masses of resistant rock that stand above the surrounding terrain). Ayers Rock (the Aboriginal name is "Uluru"), a coarse-grained arkosic sandstone, is the most famous, but nearby the Olgas ("Kata Tjuta"), comprised of massive conglomerate beds (see Photo 2-3), and Mount Conner, a gigantic flat-topped mesa with a thick quartzite caprock, are equally dramatic.

Photo 2-3 *The Olgas constitute one of the most impressive of the central Australian bornhardts. (TLM photo.)*

Almost all of the Western Plateau is arid or semiarid. Mechanical weathering, the fretting action of wind-driven rock particles, and an abundance of loose sand are much in evidence. Fully 40% of this arid landscape is veneered with aeolian sand, divided about equally between sandplain and dunefield. Extensive portions of the four principal deserts in this region—Tanami, Great Sandy, Gibson, and Great Victoria—as well as the Simpson Desert of the Central Plains Region, are covered with an array of parallel, longitudinal sand ridges that comprise vast dunefields. Indeed, these dunefields blanket about one-fourth of the entire continent. The dune pattern is a counterclockwise whorl about a long axis that lies approximately west-east along the 26th parallel of latitude. The individual dunes are long, narrow, and stabilized by some vegetation on their sides, so that they are not mobile.

The Nullarbor Plain is a vast, unusual area with an almost unbelievably flat and smooth surface. It is underlain by limestone and is characterized by an almost complete lack of surface drainage. Unlike the other dry lands of Australia, the Nullarbor Plain has no dry stream courses or intermittent lake beds.

The margin of the plateau slopes gently down to the sea in some places, as in the shoreward portion of the Great Sandy Desert. More common, however, is the presence of a rather abrupt escarpment separating the plateau proper from a fringing coastal plain. This type of development is particularly noticeable in the Nullarbor Plain in the south and along the Darling Scarp in the southwest. There is a narrow but continuous coastal plain along the western edge of Western Australia south of the Kimberleys.

Arid, with Fringes

The relatively undistinguished topography of Australia has an important, if negative, influence on the climate. The general lack of topographic diversity gives rise to broad uniformity of climate, with the result that climatic variation is gradual and transitional over most of the continent, and abrupt areal differences are very limited.

Even so, the basic fact about the Australian climate—its aridity—is in large measure occasioned by a fundamental topographic relationship, the position of the Eastern Highlands. Although we have already noted that this cordillera is neither high nor broad, its north-south orientation is athwart the prevailing easterly winds, thus effectively shutting out moist Pacific air masses from the bulk of the continent. This is by no means the only cause of Australia's expansive aridity, but it is a major one.

Apart from lack of moisture, the most notable feature of the Australian climate is its subtropicality. Its latitudinal location assures that high temperatures will be more common than low ones, that seasonal temperature differences will be minimized rather than emphasized, and that diurnal temperature fluctuations are often more meaningful than annual variations.

Thus Australia is a land of clear skies, sunshine, warmth, and little rain. Whereas such a generalization is grossly accurate, it is only an introduction to climatic veracity, an introduction that is best elaborated by consideration of the pattern of climatic regions (see Figure 2-2b).

The Arid Interior. The most extensive climatic region is the vast desert and semi-desert that occupies all of the interior and much of the west of the continent. Well over half of the country receives less than an average 15 inches (380 mm) of rainfall annually, and on account of a high rate of evaporation, desert conditions are widely prevalent. Nevertheless, Australia is remarkable for the extent rather than for the severity of its dry climates; there are no hyper-arid areas, and much of the arid zone verges on the semi-arid.

Rains are few and unpredictable, but often violent and showery when they occur. Prolonged drought is an omnipresent menace, and it is not uncommon for several years of below-normal precipitation to afflict a large portion of the region. On the other hand, flooding is occasionally experienced, when widespread thunderstorms yield short periods of heavy rain.

This is a subtropical desert region, with warm weather dominant. Summer is likely to be scorchingly hot; maximum temperatures in the 90s°F (30s°C), or higher, are to be expected for many weeks over most of the

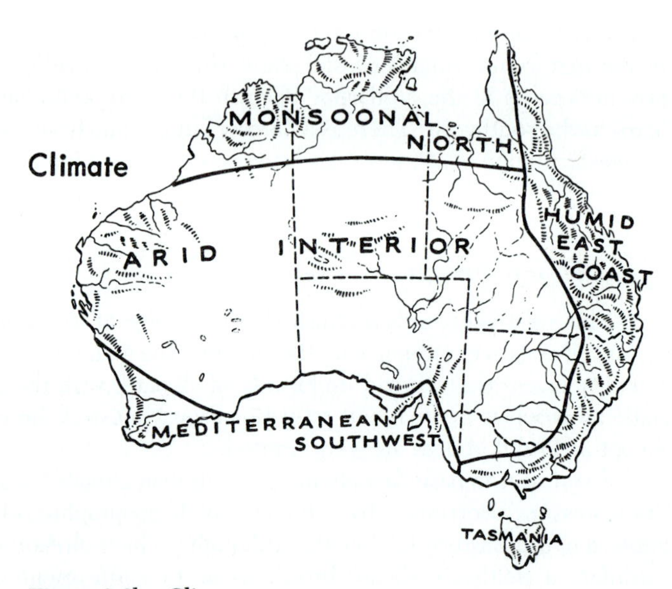

Figure 2-2b Climate

region. Rapid radiational cooling usually permits a decrease of 20 to 25° at night, but nocturnal relief is short-lived, and temperatures begin to climb again at sunrise. Gusty winds, blowing sand, and persistent flies add to the unpleasantness of summer.

Winter is characterized by long stretches of pleasant weather, with clear skies, bright sunshine, and mild temperatures (maxima typically in the 70s°F [20s°C]. Nocturnal cooling often brings midwinter minimum temperatures down into the 40s°F (5° to 10°C), and mild freezes are fairly common.

The Moonsoonal North. The climate of Australia's far north is dominated by the reversing seasonal regime of the monsoon. During four or five months of midsummer there are persistent oceanic (northerly) winds that bring abundant moisture, thunderstorms, and various tropical disturbances onto the continent. For the remainder of the year, high pressure conditions develop over the land, and outblowing (southerly) air flow more or less completely inhibits precipitation. This monsoonal regime is well developed in the Cape York Peninsula, Arnhem Land, and the Kimberleys district, and monsoonal tendencies may extend another two or three hundred miles (320 or 480 km) southward.

"Summer" is only slightly warmer than "winter" in the monsoonal north, but the former's appreciably higher humidity raises the sensible temperatures to an uncomfortable level. The summer monsoon is not a time of continual rainfall; rather there are frequent heavy showers interspersed with long periods of sunshine. Most of the monsoonal region receives from seven to 20 inches (180 to 500 mm) of rain during the wettest month, in comparison with an annual total that varies from about 25 to 60 inches (635 to 1,525 mm). The wet season is a time of decreased activity throughout the region; heavy rain combines with flat land to produce widespread conditions of flooding and persistent muddiness, which severely limit transportation.

The dry season is longer than the wet, and is characterized by mild to warm temperatures and generally pleasant weather conditions. Rain is not unknown, but is decidedly rare. As the vegetation dries out, the probability of bush fires increases, and the latter part of the dry season is notable for an almost continual pall of smoke somewhere on the horizon.

The Humid East Coast. The eastern and southeastern littoral zone of Australia is the region occupied by most of the population, and it is more than just coincidence that this is also a region with relatively abundant rainfall that is well distributed seasonally and reasonably dependable from year to year. The great latitudinal extent of the region encompasses a variation from true tropical conditions in northern Queensland to a midlatitude condition in southern Victoria. Thus, temperature variations are minimal in the north, but there are major seasonal differences in the south.

Rainfall totals vary from as little as 20 inches (500 mm) in sheltered locations to more than 100 inches (2,540 mm) in some exposed hillside sites. In general, the higher values are in the north. Overall, it can be said that precipitation ranges from adequate to super-abundant. Summer is the rainy season in the north, winter is the time of maximum precipitation in the south, and the regime is well balanced in the center. Even in this well-watered environment, however, droughts occasionally occur.

The Mediterranean Southwest. In the two southwest corners of Australia—one area centering on Perth in Western Australia and the other focusing on Adelaide in South Australia—there is a distinctive development of "Mediterranean-type" climate. This is a subtropical dry summer situation, distinguished by its inverted precipitation regime; winter is the wet season, whereas summer is almost rainless.

Indeed, summer is almost desert-like in its characteristics, except in the immediate vicinity of the coast. Temperatures are high during the day, with mid-summer maxima reaching above 100° (38°C) with some frequency. Clear skies, bright sun, and little wind movement further characterize the summer.

Winter weather is dominated by westerly air flow, which brings recurrent but irregular passage of extratropical cyclones with their associated cold fronts (warm fronts are almost unknown in Australia). Frontal passage usually results in rain, which may be brief and showery or protracted and drizzly. Annual rainfall totals in the region average between 20 and 45 inches (500 and 1,140 mm), with 20% to 35% of the total falling in the wettest winter month. Winters are not really cold, but there are long stretches of cool weather, with mid-winter maximum temperatures in the 60s°F (16 to 21° C) and minimum in the 40s°F (5° to 10° C). Snow is rare, but not unknown.

Tasmania. The island of Tasmania is far enough south to have a true mid-latitude climate, with warm summers and cold winters. The topographic complexity of the region produces considerable local climatic variations, the most widespread of which involves abundant precipitation on the western slopes and rain shadow conditions in the east. Most of the island receives at least 30 inches (760 mm) of precipitation annually, and some places experience thrice that. Winter is the season of maximum precipitation in the western part of the region; elsewhere a uniform seasonal regime is characteristic. Snow is experienced over most of the island, at least occasionally.

Water Scarce, Water Deep

To the hydrologist and hydrographer, Australia is a continent of fascination. The nature, distribution, and amount of surface and sub-surface water has many unusual facets and is comparable, even in its basic aspects, to that of no other continent. In essence, Australia has the smallest supply of surface water and the most remarkable conditions of underground water of any of the settled continents.

Hydrography. The basic fact of Australian hydrography is the scarcity of water. The country lacks the high, young mountain ranges that are found in other parts of the world, so there are no permanent snowfields or glaciers to sustain river flow. Furthermore, orographical precipitation possibilities are limited, to say the least. As a generalization, it can be stated that wherever the annual precipitation is less than about 20 to 25 inches (500 to 635 mm), no permanent streamflow can be maintained. Under these conditions, streams are either ephemeral (flowing during the wet season or after rains only) or exotic (sustained by waters that enter the drainage system from wetter areas). Thus, nearly three-quarters of the country—all but the northern, eastern, southeastern, and southwestern coasts—is largely without permanent streamflow. Only in the two smallest states, Tasmania and Victoria, do the majority of the principal streams run continuously.

Most of the important Australian rivers have their headwaters in the Eastern Highlands, which serve as the east-west continental drainage divide. In many areas, however, this divide is not at all sharp, and the streams originate in grassy flats or tablelands where the watershed is not clearly defined.

The principal drainage basins are characterized in the following list:

1. The *Carpentaria Basin* is a classic example of a centripetal drainage system, with rivers converging toward the Gulf of Carpentaria from dis-

tant, slightly elevated uplands to the east, south, and west. Total annual streamflow in the basin is very large, but strictly seasonal. Most of the rivers, with the conspicuous exception of the Gregory and some from the northern Cape York Peninsula, flow only during and immediately after the period of summer monsoon; during the remainder of the year they are mostly dry. The entire drainage basin reflects the importance of monsoonal rainfall. Almost all the rivers cross a very flat coastal plain in their lower courses, and are characterized by much distributary development and mutual discharge into neighboring rivers. Most flow into extensive mangrove swamps on the edge of the Gulf, and many have no outlet to the Gulf proper. The most notable rivers in the Carpentaria Basin are the Roper from the west; the Gregory, Leichhardt, and Flinders from the south; and the Gilbert and Mitchell from the east. This last-mentioned stream may yield the greatest total annual discharge of any river in Australia, even exceeding the Murray, although accurate statistics to verify this claim are not available.

2. The *Mainland Pacific Slope* encompasses a large number of relatively short streams that carry, on balance, a considerable annual flow. Most of the rivers are structurally similar in that they rise fairly near the coast and then flow latitudinally (north or south) in their upper courses for some distance at a modest gradient before turning more or less abruptly seaward. The monsoonal influence is strong in northern Queensland, and the rivers there have a pronounced flow maxima in summer. Farther south, however, winter is the wetter season, so that most Pacific slope rivers of New South Wales and Victoria carry more water in that season. The greatest annual discharges are from the Tully, Herbert, Burdekin (second only to the Murray in measured annual flow), Fitzroy, Burnett, Richmond, Clarence, Hunter, and Snowy.
3. The drainage systems of *Tasmania* consist of a number of short rivers that have considerable discharge, due to relatively heavy, year-round precipitation. Many of the streams have broad estuaries at their mouths. Tasmania's largest flow is in the Derwent system, which drains the southeastern part of the island. Other major rivers are the Huon in the south, the South Esk and Mersey in the north, and almost every river on the west coast.
4. The *River Murray system* is by far the most important drainage in Australia, and also the most complex. Total annual discharge from the Murray is about 8 million acre-feet (in comparison to 474 million acre-feet from the Mississippi), but half again that much is withdrawn for irrigation along its course. Except in its high headwaters, the Murray has a remarkably gentle gradient; from Albury to the mouth (more than 1,200 miles [1,900 km]) the gradient is less than nine inches to the mile (14 cm per km), and over more than half that distance it is less than 3 inches to the mile (5 cm per km). As might be expected, then, there is considerable development of distributaries and marshes with maze-like channels, although in some

areas the stream has a narrow incised valley in a broad floodplain. Major tributaries of the Murray include the Murrumbidgee, which drains the northern slopes of the Snowy Mountains; the Goulburn, from Victoria; and the Darling, a lengthy river with a very erratic flow regime.

5. The drainage systems of *Southwestern Australia* are characterized by a number of relatively short streams that have their upper courses in wide, flattish valleys on the plateau surface and then descend onto the coastal plain and into the sea. Many of the streams empty into lagoons because they are not strong enough to maintain an open entry into the ocean. Their flow regime is often intermittent, with winter as time of maximum discharge.
6. The *Western* drainage systems contain a few long rivers that are almost entirely intermittent in their flow. They have the characteristics of desert streams, normally dry but carrying overflow flood water after the infrequent rains. Evaporation and sinking into the sand take much of the water that would otherwise flow into the Indian Ocean.
7. The *Timor Sea* drainage systems have markedly monsoonal regimes. Extensive floods are common in summer, but during the dry season most of the rivers either dry up completely or are reduced to irregular chains of billabongs (pools or waterholes). The great rivers of this region are the Fitzroy (which drains the southern Kimberleys), the Ord (which drains the eastern Kimberleys), and the Victoria and Daly (in the Northern Territory).
8. The *Lake Eyre Basin* is one of the largest areas of internal drainage in the world. The focus of convergence of this system is a series of interconnected lakes in South Australia, of which Lake Eyre is by far the largest. The lakes are playas, and are normally dry with a heavily salt-encrusted surface of silt and sand. On rare occasions they become water-filled, but never deeply. The principal drainage into the lakes comes from western Queensland via Cooper's Creek and the Diamantina and Georgina rivers. Their waterways make up the famous Channel Country. This region is dry most of the time, but occasionally after a heavy rain an enormous area will be flooded, bringing about a rapid growth of herbage that is much desired for livestock pasturage.

There are other large areas of internal drainage in Australia, as in the Wimmera district of northwestern Victoria or the Bulloo system of south-central Queensland, but the most extensive "drainage" region of all is one that is largely without any sort of surface waterways. This region extends from the Great Sandy Desert in the north to the Amadeus Basin and the Great Victoria Desert in the east to the south edge of the Nullarbor Plain in the south, and occupies about one million acres (400,000 ha) of land. A few intermittent streams are found, but for the most part this region simply has no surface drainage at all.

Underground Water. In contrast to the scarce surface water, Australia's underground water resources are unusually plentiful. Their quality, unfortunately, does not match their quantity. Actually, the *ground water* supply (water in the saturated zone between the water table and the uppermost impervious layer) is not unusual; indeed, it is distributed about as one might expect in an arid continent. The resources of *confined water* (occupying deeper aquifers under pressure between impervious strata), on the other hand, are quite widespread. The confined water is largely *artesian* (under sufficient pressure to rise to the surface when tapped), though some is *subartesian* (under less pressure, and so will rise only part way to the surface when tapped).

The most famous and by far the most extensive of the underground water supplies is the Great Artesian Basin, which underlies two-thirds of Queensland, much of New South Wales and South Australia, and part of the Northern Territory (see Figure 2-2c). This is the largest artesian basin in the world, with several thousand flowing bores (wells) coming from three different aquifers (see Photo 2-4). A number of problems inhibit maximum use of the water, however, particularly depth, temperature, and salinity. In some places the water source is more than 7,000 feet (2,100 m) below the surface, significantly increasing the cost of well drilling. Normally the water is quite hot, occasionally with a temperature that rises as much as 1° F. for every ten feet (3 m) of depth; sometimes the water must be run in an open ditch on the surface for half a mile (0.8 km) before it cools sufficiently for cattle to drink it. The dissolved mineral content in the water is usually

Figure 2-2c Artesian Basins

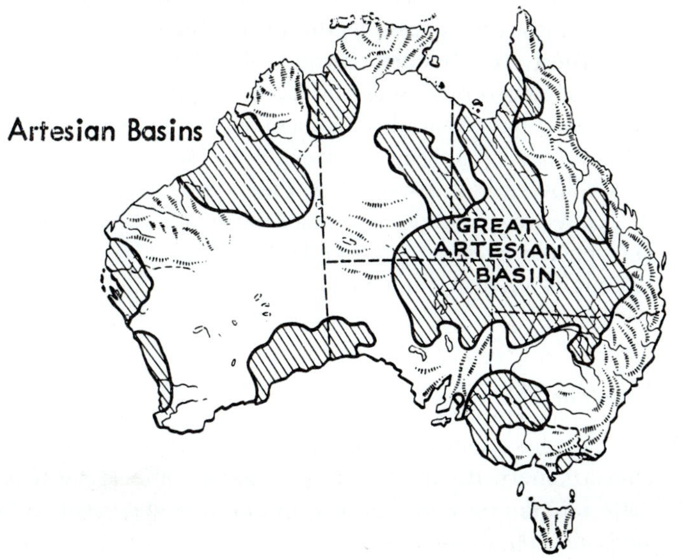

Photo 2-4 *A typical artesian bore in the Great Artesian Basin. This is the Pilliga bore in northern New South Wales. (TLM photo.)*

so high as to make the liquid unsuitable for human use or irrigation purposes, although it is satisfactory for stock watering.

There are other artesian and subartesian basins in Australia, but none nearly as large or as significant as the Great Artesian Basin. In some cases these other basins have better quality water, but for the most part they suffer from the same handicaps as does the Great Artesian Basin.

Subsurface water supplies, then, while vast in extent are poor in quality. They are extremely important to pastoral development, but should not be expected to contribute significantly toward "opening up" the country to farming or urban settlement.

The Skin of the Earth

We have seen that Australia is a land of unusual structure, topography, climate, and hydrography; thus it should come as no surprise that the soil pattern of the country differs in several important aspects from the generalized continental pattern that is found over most of the world. The appropriate generalization is that the humid soils of Australia are quite similar to their counterparts on other continents, but the Australian arid soils are much less comparable. Presumably, a soil that develops under conditions of considerable leaching will turn out in a "normal" or predictable fashion; that is, the climatic influence will predominate. However, for soils of drier areas the relief or parent material will be a much more significant influence on its development. Thus the ancient, quiescent, and subdued nature of the structure and topography of this continent have left an unusual brand on many of its dry land soils.

As a corollary of this, there is a considerable proportion of ancient and generally impoverished soils in Australia. With a few exceptions, the soils of the continent have a low level of natural fertility, owing to lack of moisture and an inadequate supply of plant nutrients. There is a singular scarcity of phosphorus and nitrogen; only potassium and calcium among the important chemical elements seem to be present in generally adequate quantities. Excess salinity is a major characteristic of many soils and thus a significant land-use problem.

Australian pedologists and agronomists have done a great deal of work on soil problems, particularly on the chemistry of trace elements. The application of their results has meant a notable difference in agricultural and pastoral productivity in many areas; so much so that fertilization is a major factor in Australian consciousness. Such terms as "superphosphate" are now household words (not inappropriately in a country where more than 2,500,000 tons (2,250,000 t) of "super" are applied to the soil annually).

Soil Patterns. Approximately 50 great soil groups have been identified in Australia, and their pattern of distribution is quite complex. In the broadest sense there is a congruency between gross climatic and soil patterns, so that the effect of aridity is clear from any soil distribution map. The reader is reminded, however, that only the most general of soil associations can be shown on a small-scale map like Figure 2-2d; thus, more detailed pedogeographic scrutiny of any part of the continent would demonstrate many variations from the broad pattern.

Humid land soils are restricted largely to east coastal locations, from Cape York to Tasmania, with a small strip along the southwest coast of Western Australia. These are mostly podzolic-type soils that are only partially analogous to types in North America or Russia. The surface horizons are usually grey, the soils tend to be acidic, and fertility is low to moderate. In a few scattered areas are found red loams called krasnozems that have been developed on basaltic (or similar) bedrock that provides plenty of ferric oxide to keep the soil well flocculated; these are some of the most fertile of Australia's soils.

Subhumid soils are found in a more inland crescent in the east, particularly in New South Wales and southern Queensland, though some also occur in a similar position in Western Australia. A large proportion of the soils in this group consists of blackearths that are similar to typical chernozems, but with a structure that is more likely to be granular or cloddy than crumby

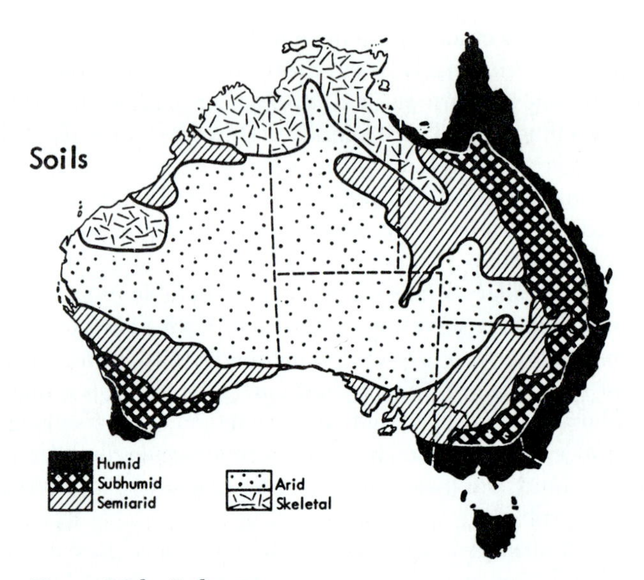

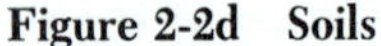

Figure 2-2d Soils

and a more limited humus contents. Red-brownearths are also widespread in the subhumid zone; they have moderate fertility and can become quite productive with careful management.

Semiarid soils are spread in a ring that goes nearly four-fifths around the heart of the continent. These are quite varied grey and brown soils, often solonized, that are of moderate to low fertility.

Arid land soils include a considerable variety of types and are generally distributed over nearly half the continent. Their color varies, but shades of red and red-brown are predominant. These soils posses a number of unusual characteristics; for example, some of them are distinctly acidic in reaction, in contrast to the alkaline reaction normally expected of desert soils. On the whole the soils of the arid zone can be said to be less fertile than typical desert soils of other continents. Furthermore, there are extensive areas where no true soils have developed, and either gibber plains or largely unanchored sand ridges dominate the surface of the land.

Skeletal soils are widespread, especially in the monsoonal lands of the north, particularly in Western Australia and the Northern Territory. These are poorly developed and thin, and are mostly associated with ranges and tablelands.

A Sclerophyllous World

The natural vegetation of Australia is another environmental facet that exhibits the uniqueness of the land Down Under. Such development is thought to be primarily the result of isolation. The Australian tectonic plate broke away from Gondwanaland (the southern supercontinent) about 100 million years ago, drifting northeasterly and carrying with it the ancestors of much of its present biota, both plant and animal. The separation of the continents allowed Australia's isolated populations to evolve and diversify without interference from immigrant species. The numerous ecological gaps that appeared were filled by evolution of homegrown forms rather than by immigration from outside. The Gondwanaland heritage included a rich diversity of both plants and animals living in rainforest habitats. From this there evolved a new suite of forms adapted to coping with aridity, impoverished soils, and wildfires. Nowhere else in the world did this happen under conditions of isolation by sea barriers. Thus, only in Australia did such an unusual array of species develop and survive to the present.

For several tens of millions of years the Australian plate drifted in total isolation, until it finally began a protracted collision with the island chain to the north. Only in the last few million years, then, has there been "contamination" of the Gondwana heritage. As the collision event continued, inter-island distances decreased until island-hopping provided a limited interchange of biota between Australia and the biotic richness of Southeast Asia. Thus the native flora of Australia is both diverse and specialized. It is estimated that there are some 25,000 species of plants on the continent, and 80% of these are endemic (i.e., native *only* to Australia). Indeed, about 30% of Australia's plant genera are endemic. This high rate of endemism accounts for the distinctively Australian character of the flora.

The most notable gross feature of the contemporary Australian floristic pattern is the scarcity of dense forest. Nearly all of the continent is vegetated with grassland, shrubland, or relatively open woodland. Furthermore, it is the nature of the dominant tree species (eucalyptus) that their leaf density is relatively sparse and the long narrow leaves hang down inertly, a position that amplifies and emphasizes the openness of the forest or woodland. Eucalypts are classed as evergreen trees because they do not experience a seasonal leaf fall, but their soft grayish pigmentation makes it seem logical to call them "evergrays."

Two plant genera are overwhelmingly dominant. The forests, woodlands, and part of the shrublands are largely composed of species of eucalypts (*Eucalyptus* spp.), usually referred to as "gums," whereas much of the rest of the country—shrublands and grasslands—features acacias (*Acacia* spp.), often called "wattles." It should be noted that there are about 450 species of eucalyptus and 900 species of acacias in Australia.

As is to be expected in an arid continent, much of the Australian flora exhibits pronounced xerophytic (drought-resistant) characteristics, such as deep taproots and great root density to seek moisture in the ground, small and hard leaf surfaces to inhibit transpiration of water, and shiny plant surfaces to reflect rather than absorb solar insolation. Even so, there is a complete absence of native cacti and similar succulent forms that are so common in North American desert areas.

Major Vegetational Associations. The broad pattern of natural vegetation in Australia (see Figure 2-2e) shows a narrow forest zone (mostly open, sclerophyllous forest, but with scattered patches of dense rain forest) along the east coast and in the extreme southwest. Inland from the forest, and encircling most of the continent, is a zone of woodland, mostly quite open. Farther inland is a complete ring of mixed grassland and shrubland. In the core of the continent, but displaced toward the west, is an extensive area dominated by true desert, often with more bare ground than vegetation of any kind.

Rainforest occurs in small, discontinuous patches scattered along the entire east coast, from Cape York to southern New South Wales, although the principal concentrations are in Queensland. This is essentially tropical rainforest, which grows only in response to abundant precipitation, typically at least 60 inches (1,525 mm) annually, and protection from the drying effect of day-long sun; hence, occurrences are limited mostly to south- and east-facing slopes. It is an evergreen forest comprised of a great variety of species, which are mostly Malaysian in origin (and in which eucalyptus is almost entirely absent). There is normally a dense, interlacing canopy, beneath which there may be one or two other layers acting as discontinuous subcanopies. Lianas and epiphytes are characteristically interspersed. A large area of temperate rainforest, characterized by smaller trees and a lack of parasitic growth, occurs in southwestern Tasmania, and there are smaller patches in Victoria and New South Wales.

Much more extensive is the *sclerophyll forest* association, which is the dominant vegetation within 100 miles (160 km) of the coast of New South Wales and most of Victoria, and also occupies much of Tasmania and the southwestern corner of Western Australia (see Photo 2-5). This is an essen-

Figure 2-2e Natural Vegetation

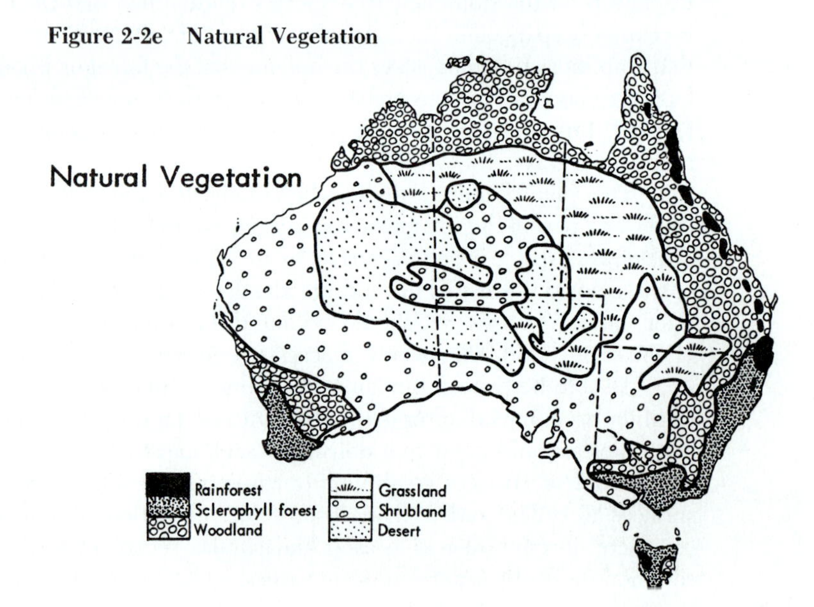

Photo 2-5 *A dense karri forest in southwestern Western Australia, near Walpole. (TLM photo.)*

tially eucalyptus forest community made up of medium to tall trees with crowns that form a relatively complete and interlacing canopy. Undergrowth may be dense, but is usually relatively open.

More extensive still is the *woodland* association, comprised of an open, tree-dominated community. This is the principal plant association of the north, of most of the eastern third of Queensland, and of a large area inland from the forest zone of New South Wales, Victoria, and Western Australia. The trees are of varying height and their branches do not interlace into a canopy; thus, there is a distinctly open aspect to the growth. There may or may not be undergrowth, but a grassy understory is typical.

The Australian *shrublands* generally occupy the southern half of the continent except where there is forest, woodland, or desert. The term "shrubland" refers to a considerable variety of shrubform associations that may grow either densely or openly, may or may not be considerably intermingled with grasses, and may be extensive or discontinuous in pattern. In some cases the dominant species take a tree form, but they are never very tall. A special shrubland association is the *mallee*, a name applied to country in which any of a number of distinctively shaped eucalypts grow; they are low trees or tall shrubs with many stems branching from a subsurface rootstock (lignotuber). Also notable are the extensive shrubby steppelands dominated by a bluebush-saltbush (*Kochia* spp.- *Atriplex* spp.) association, which look much like the sagebrush plains of the western United States but are valuable for livestock browsing.

Grassland associations cover much of the northern and eastern interior of the country. The grasses are often interspersed with scattered trees or varied shrubs, particularly near the moister margins. Most of the grassland areas are covered with discrete tussocks rather than a continuous sod; thus "tussock grassland" is a proper term of widespread application. Two extensive grassland types of special interest are the Mitchell grass downs (excellent grazing lands dominated by *Astrebla* spp.) and spinifex associations (where great spiky clumps of *Triodia* spp. dominate).

Much of the interior of the country, extending well into the northwest, is true *desert*, and hence very sparsely vegetated. A scattering of xerophytic grasses (mostly annuals) and shrubs is found, and low trees sometimes grow in the dry watercourses. Spinifex is a common plant in the deserts (see Photo 2-6).

Photo 2-6 *Rounded hummocks of spiky spinifex comprise the dominant vegetation over nearly one-fourth of the continent. This scene is near Windorah in southwestern Queensland. (TLM photo.)*

Human Influence on the Flora. The preceding discussion refers to natural vegetation, much of which has been modified by human activities and influences, starting some 50,000 years in the past with the arrival of the first Aboriginal people, and being severely accelerated with the arrival of European settlers two centuries ago.

There is no evidence that Aborigines brought plants with them when they came to Australia. The principal impact of their habitation was a significant increase in the incidence of fire, which they used as a tool for the deliberate management of vegetation.

Europeans relied less on the use of fire, but deliberately and accidentally introduced many plant and animal species. By clearing large areas of natural vegetation, they modified the habitat of the original flora, encouraging invasions by alien plants. In many areas, the original flora has completely disappeared, as a result of clearing the land for farming or some other intensive use, or from pastoral activities.

The principal form of land use in Australia is pastoralism, and most stockmen are eager to improve their grazing conditions by any means. "Improvement" of the land often involves the elimination of trees to make more room for grasses; this is frequently accomplished by "ringbarking," or girdling the tree. The tree dies sooner or later, and the dead trunk usually is left standing for years, until it eventually blows over.

Exotic grasses are sometimes introduced to upgrade the pasturage, so that vast acreages of Australia are now sown to such species as crested wheat grass from the Ukraine and bluegrass from the United States. Many other exotic plants have been introduced into Australia, and it is estimated that about 15% of the total Australian flora consists of introduced species.

Some of the plant introductions have had disastrous consequences. For example, during the 1920s, prickly-pear cactus (*Opuntia* sp.) was brought from North America into Queensland and northern New South Wales, and quickly infested more than 50,000,000 acres (20,000,000 ha.) before it was essentially eliminated. Indeed, it is reported that "all of the 18 world's worst weeds are present in Australia—and not one of them is native."[1] Much more satisfying has been the introduction of various species of coniferous trees in an effort to establish a softwood lumber industry. Every state now has prospering plantations of exotic pines, primarily *Pinus radiata* from California.

Another destructive impact on the Australian flora by Europeans is the accidental introduction of the cinnamon fungus, a parasite that causes dieback (slow deterioration and death) of native trees and shrubs. This is a major problem in the southwestern corner of the continent, as well as in parts of southeastern Australia and Tasmania. Quarantine measures have slowed down the spread of the fungus, but there are as yet no known ways to prevent or cure the problem.

[1]Jeremy Smith (ed.), *The Unique Continent* (St. Lucia: University of Queensland Press, 1992), p. 159.

A Faunal Asylum

Of all the Australian environmental elements, none is so remarkable as its fauna; its assemblage of terrestrial animal life is completely without parallel in other parts of the world, and even its bird life is significantly different from that of other continents. When the Gondwanaland supercontinent became fragmented, the drifting tectonic plates each became separate evolutionary units. Some were close enough to others, at least in the early stages, that there were overlapping influences. However, during most of the interval of mammalian evolution, the Australian plate was sufficiently removed from other plates that its evolutionary development was separate and distinct. It became a sort of "faunal asylum" in which rare and vulnerable species were able to survive because of the absence of competitors and predators.

Thus the mammalian fauna of Australia is bizarre in the extreme. Other major faunal groups—birds, reptiles, amphibians, fishes, invertebrates—are not so distinctive, presumably because much of their evolution took place earlier, and was more influenced by evolution on other continental plates, especially Asia and India.

Placental mammals, the dominant animals of other continents, are limited and inconspicuous in Australia. Although more than 100 placental species are recognized, they are all bats, rats, or mice. The "normal" placental groups—ungulates, primates, felines, canids, mustellids, and others—are totally absent.

The characteristic mammals of Australia are marsupials, relatively primitive types that give birth to partially developed, almost embryonic young, which develop after birth for a long period in the mother's pouch. Marsupials lack a well-developed womb and have no placenta, so cannot give their young a long safe period of internal growth. Marsupial embryos are attached to the wall of the uterus, and are nourished by osmosis. As the embryo begins to develop into a complex organism, osmosis becomes inadequate to provide nourishment, so the tiny creature is born in a partially developed state. The gestation period is remarkably brief, lasting only 8 to 40 days, depending on the species. The newborn baby is unformed in almost every facet except its head and forelimbs; the latter enabling it to make the hazardous journey from the birth canal to the pouch, where it attaches itself inseparably to a nipple for several weeks or months. The baby marsupial spends from two to five times as long in the pouch as it does in the womb.

In the absence of placental predators and herbivorous competitors, marsupials dominated Australia over a very long period of time. They advanced to a stage of development unknown on other continents. The variety of species that has developed and the diversity of ecological niches that these species have filled is extraordinary. In the course of their persistent radiation, they have converged to a remarkable degree, ranging from treetop herbivores to nocturnal predators.

Most mammalogists believe that the original cargo of mammals carried by the drifting Australian plate consisted only of marsupials and monotremes; these vulnerable species had "the run of the ship" for a very long time. Only in the last five or 10 million years did placental mammals begin to come aboard. The principal unfilled niches in the Australian ecosystem apparently were those occupied elsewhere by certain placental rodents that are small seed and vegetation eaters with specialized jaws and teeth adapted to a hard, woody diet. Eventually then, placental rats and mice from Southeast Asia came to Australia, where they developed distinctively from their Asian forebears. Bats came to Australia during the same time-frame. There are now some 70 species of rats and mice in Australia, and about 50 species of bats.

The Australian marsupials include a dozen Recent families, embodying more than 120 living species. The majority are herbivorous types, including the well-known macropods (kangaroos and wallabies), of which there are about four dozen species (see Photo 2-7); the numerous rat kangaroos; the bulky wombats, which are badger-like diggers; tree-dwelling phalangers and possums, numbering some 40 species; and the single species of koala. There are also a number of carnivorous marsupials, mostly small, including thirty species of marsupial "mice," a number of marsupial "moles," various "cats" and "devils," the presumably extinct Tasmanian "wolf," and a group of num-

Photo 2-7 *A western gray kangaroo near Cape Le Grand on the south coast of Western Australia. (TLM photo.)*

Photo 2-8 *The echidna, or spiny anteater, is one of only two types of monotremes in existence. This Victorian scene is on the slope of Mt. Donna Buang east of Melbourne. (TLM photo.)*

bats (anteaters). Finally, there are about twenty species of omnivorous marsupials, called bandicoots. These various marsupials range widely over Australia, but are found nowhere else in the world except New Guinea and some associated islands.

As an acme of primitive Mammalia, Australia is also the home of the world's only contemporary monotremes, the duck-bill platypus (*Ornithorhynchus anatinus*) and the spiny anteater (*Tachyglossus aculeatus*) (see Photo 2-8). These are the only egg-laying mammals to be found anywhere, except New Guinea.

Australia also has a numerous and varied reptilian fauna. Although turtles and tortoises are limited, there are 450 species of lizards, including some large ones. Two species of crocodile inhabit the northern rivers and lagoons. Snakes are numerous both in quantity and variety.[2]

[2]Venomous creatures are relatively abundant. A story in *International Wildlife* (March-April 1988) enumerated the "10 most venomous animals in the world"; seven are found in Australia—box jellyfish, beaked sea snake, blue-ringed octopus, stonefish, funnel-web spider, taipan, and eastern brown snake.

Bird life is exceedingly varied. Some 720 species are known in Australia, of which 570 breed on the continent and more than 325 are endemic (see Photo 2-9). Especially notable are the psittacine (parrot-like) types, which occur in greater diversity than on any other continent. Two of the world's largest land birds, the emu and the cassowary, both flightless, are also notable.

Lesser forms of animal life are more limited. Amphibians, for example, are sparse, represented by only a few species of frogs and toads. Freshwater fishes are also restricted by the general paucity of surface waters. Insects and associated arthropods, on the other hand, are quite abundant; particularly notable are flies, white ants (termites), ants, butterflies, and mosquitoes.

Natural Hazards

Rural Australians, and sometimes their urban counterparts, are afflicted from time to time by four notable natural hazards—droughts, floods, cyclones, and bushfires.

Photo 2-9 *The kookaburra is a large kingfisher that is often referred to as the symbol of Australia. The locale of this photo is along the Margaret River in southwestern Western Australia. (TLM photo.)*

Focus Box: Exotic Wildlife

In addition to native forms of wildlife, a conspicuous element in the present faunal complement of Australia is comprised of foreign species that have been introduced to the continent, by accident or design, and have become established in the wild. Two factors account for the significance of non-native wildlife in the Austral continent:

1. The native animals are generally limited, specialized, unaggressive, and vulnerable, making it relatively easy for an exotic to become ensconced, once it is introduced.
2. European settlers, particularly in the 19th century, were often eager to introduce animals from the "old country" to "improve" upon the sparse and unusual native fauna.

The earliest of the exotic introductions was the dingo, which apparently was brought to Australia by Aboriginal migrants from Southeast Asia, and has been a well-established member of the fauna for many centuries. The most notable recent introduction was that of the European rabbit, whose spread from an initial advent of 24 animals near Melbourne in 1859 to a half-continent plague within 50 years is the classic scare story of all mammalian importations. The diffusion of the European fox over Australia was even more expansive (the rabbit never moved into the northern third of the continent, whereas the fox has spread to almost every corner of the land).

In many ways, the proliferation of feral livestock in Australia is even more spectacular than the story of either the rabbit or the fox. One of the most striking discrepancies between the Australian wildlife assemblage and that of other continents is the total lack of native ungulate (hoofed animals) species. Ungulates are generally large in size, gregarious in habits, and widespread in distribution; thus their presence is normally conspicuous in the landscape, and their absence from Australia is a notable zoogeographical oddity.

All six of the common mid-latitude barnyard animals (horses, donkeys, cattle, sheep, goats, and pigs) have been bred as domesticated livestock in considerable numbers over most of the settled parts of the continent. In addition, two more specialized, subtropical livestock varieties—dromedaries and water buffaloes—were brought to Australia in limited numbers. Seven of these species (all except sheep) have on occasion escaped from confinement or have deliberately been turned loose, and have established free-ranging populations on a sizable scale in various parts of the country, thus becoming *feral* in the true sense of the term.

At the present time, Australia has more feral horses, donkeys, cattle, goats, water buffaloes, and camels than any other country in the world, and ranks second (to the United States) in numbers of feral pigs. The total number of these animals in the country varies from year to year, but is in the general vicinity of three million individuals. There are also uncountable numbers of feral dogs and cats.

The presence of such a vast population of exotic wildlife is almost universally deplored. Most objections are on economic grounds, particularly by pastoralists. Dingoes are effective predators on sheep; rabbits and feral ungulates inhibit pastoral operations, primarily by consuming feed and water that graziers want for their livestock but also by a variety of other transgressions; foxes, feral dogs, and feral cats are devastating predators upon such native wildlife as ground-nesting birds, lizards, and vulnerable marsupials.

Consequently, enormous efforts have been, and continue to be, expended to control these exotics by such measures as poisoning, trapping, shooting, and the building of barrier fences. The implacable opposition of the pastoral industry to these animals is thoroughly understandable from an economic point of view. And even when considered from the loftier ethic of the integrity of the continental ecosystem, it would seem the better part of wisdom to beware of the persistence of such exotic animals in the wild. One of the broad, usually ignored, lessons of history is that humankind's tampering with the biota usually has unsatisfactory results.

Drought. Australia is a country that is one-half desert and one-fourth semidesert; thus there is little precipitation over most of the continent. It is well documented that arid regions receive precipitation that is not only scarce but also unreliable; in other words, there is often wide fluctuation from the average in any given year. Thus an arid continent like Australia can be expected to experience below-normal rainfall with considerable frequency. Indeed, drought is a recurrent phenomenon in many parts of the country, and it is not unusual for large areas to endure five or six consecutive years of drought conditions.

Although the native Australian plant and animal life is adapted to aridity, and the most prominent breeds of sheep (Merino) and beef cattle (Hereford and Shorthorn) can survive conditions of considerable harshness, any prolonged drought brings severe hardship to rural Australians. Many head of livestock literally starve to death, and many others must be slaughtered prematurely to reduce the herds to fit the diminished forage. It is little wonder that drought is the primary concern of most Australian pastoralists.

Farmers, too, can be seriously impacted by drought. The major grains (wheat, barley, oats), in particular, experience dwindling yields in dry times. Irrigation has become increasingly widespread in Australia, in part to serve as a buffer against drought. Thus, water storage and distribution projects are widely used by farmers in the watershed of the River Murray, and to a lesser extent in many other drainage basins.

Flood. One of the major ironies of the Australian environment is the fact that flooding is also a significant natural hazard on this the driest of the inhabited continents. Most flood problems result from intense localized rainstorms along the humid east coast of the country, where narrow valleys and flat floodplains are susceptible to concentrated runoff. In addition, much of the monsoonal north of the continent experiences annual flooding during the heavy summer rains. And even in the arid interior there are occasional episodes of flooding that are particularly stressful because of their unexpectedness.

Cyclone. Various kinds of storms affect Australia, but by far the most devastating are tropical cyclones (identical to *hurricanes* in the United States). These intense low pressure systems strike the northeastern (Queensland), northern (Northern Territory), and northwestern (Western Australia) coasts from three to six times a year, always in the Southern Hemisphere summer. Their mighty winds and tempestuous rains can do immense damage to coastal areas; for example, a cyclone hit Darwin (administrative center for the Northern Territory) on Christmas morning of 1974, killing 60 people and destroying or damaging almost every building in the urban area.

It should be noted, however, that destruction and tragedy are not the only legacy of cyclones. Cyclone-induced rainfall is often a critical source of moisture for the parched Outback. Although devastation may result in the immediate path of the storm, a much more extensive area may be nurtured by the life-giving rains.

Bushfire. No continent is as susceptible to fire as Australia. In no other continent is such a high proportion of its area and vegetation burned so easily and so often. The inter-related reasons for this situation include soil

Photo 2-10 *Bushfires are commonplace during the dry season in the Top End of the continent. This blaze is in Kakadu National Park. (TLM photo.)*

infertility, the widespread extent of dry weather, and a long history of deliberate burning by Aborigines. These conditions predisposed the vegetation toward more fire-prone types. As a result, wildfires (usually called "bushfires" in Australia) are to be anticipated with considerable frequency over much of the continent. The monsoonal north is especially prone to burning; during the long dry season bushfire smoke is almost constantly on the horizon (see Photo 2-10). Even the arid portions of the Outback, with sparse but flammable vegetation, experience frequent fires. The more densely vegetated regions—the forests and woodlands of the east and southeast—also are susceptible to fire, particularly during periods of drought.

Bushfires may cause devastation and hardship on the local flora and fauna, as well as on crops, pastures, and livestock. Moreover, in some cases out-of-control fires, usually exacerbated by drought conditions and high winds, sweep through settlements, leaving death and destruction in their wake. For example, the infamous "Ash Wednesday" fires of February 1983 caused 71 human deaths and hundreds of millions of dollars in property damage in Victoria and South Australia.

It should be noted, however, that a bushfire is a catastrophic event, but does not necessarily create a permanent catastrophe. Many plants grow better after fires, most eucalypts regenerate very rapidly unless they are completely destroyed (they have dormant buds beneath the bark that are triggered to develop only after the crown has been severely damaged), and the ecosystem may be better off in the long run for experiencing fairly frequent burning.

CHAPTER THREE

Peopling the Austral Continent

The prehistory of Australia is only imperfectly understood; archaeological evidence is limited, but more is being discovered and interpreted each year. It is absolutely clear, however, that the Aboriginal occupance of the continent has been in place for millennia; the first Aboriginal occupants arrived at least 50,000 years ago, and this date may be pushed back significantly as more evidence continues to be unearthed. It is instructive to contemplate that the history of Australia encompasses perhaps seven European generations, but the Aboriginal prehistory comprises at least 2,000 generations!

The First Australians

The unique and remote Australian environment provided a home for a distinctive society of prehistoric people usually referred to simply as *Australian Aborigines.*[1] This society, with its Paleolithic culture, was well adjusted to the harsh realities of life on an arid continent.

Their genesis is indistinct, but it is thought that they originated in the general area of the East Indies and made their way to Australia by what were probably the first sea voyages anywhere in the world. Through the centuries they spread across and occupied all parts of the Austral continent, at varying densities depending on local environmental conditions.

Mostly they led a semi-nomadic existence, but their movements were not helter-skelter, as they had well understood territorial limits beyond

[1] In the last few years, many Aboriginal people, particularly in Victoria and southern New South Wales have preferred to be called by the self-ascribed name *Koori*. The name *Murri* is also used in northern New South Wales and southern Queensland. Neither name is as yet universally accepted by Aboriginal people.

which they strayed only under unusual circumstances. It is thought that there were 500 to 600 tribes or tribelets, and each tribe recognized the territoriality of others. Within each tribe there were usually several clans of a few dozen people each, and normally each clan had as its nucleus a small band or descent-group.

Aboriginal people had no permanent homes, often living for months at a time without any sort of constructed shelter other than a brush windbreak. In some areas of greater resources, however, stone dwellings with turf or brush roofs might be constructed for a long stay.

Their way of life depended basically upon hunting, gathering, and fishing. The bushcraft of the Aborigines (both male and female) has long been a subject of awe. They were remarkably adept as practical naturalists and trackers.

It has long been believed that Aborigines had no domesticated plants, but some recent evidence indicates that yams may have been cultivated in some areas. Certainly there was no formal agriculture as we know it, but the gathering and distribution of seeds and the management of natural vegetation by "firestick farming" indicates some degree of plant husbandry. And plant foods often required complex processing—washing, pounding, straining, and leaching—because many of them contained poisons that had to be removed before eating. They had no domesticated animals except the dog, which presumably is the source of the dingo. The normal division of labor called for men to hunt and women and children to gather; successful hunting meant good eating, but day-to-day subsistence was more directly dependent upon the success of the women as they gathered and dug seeds, roots, berries, grubs, lizards, and other Outback delicacies. Their most widespread management practice was burning. Fires were used to drive animals out of concealment for hunting. Fire regimes were developed that kept the vegetation more open and more easily traversed. Plants with edible tubers became more common, and fresh green grass growing on burned ground attracted animals to places where they could be hunted more easily.

In coastal areas and along some rivers, fishing was important. Hooks, spears, traps, nets, and poisons were used in fishing, and in some locales (notably western Victoria) large weirs were constructed to trap fish and lengthy canals were built to snare migrating eels.

They had few material possessions, apparently not wanting to be encumbered in their nomadic wanderings. Captain James Cook, perhaps the first European to have significant contact with Aborigines, remarked in wonder at their lack of interest in the gifts he gave them, a reaction that was in distinct contrast to his experience with any other native group.

> "They covet not magnificent houses, household stuff, etc.; they live in a warm and fine climate, and enjoy every wholesome air, so that they have very little need of clothing; and this they seem to be fully sensible of, for many to whom we gave cloth, etc., left it carelessly upon the sea beach and in the woods, as a thing they had no manner

> of use for; in short, they seemed to set no value upon anything we gave them, nor would they ever part with anything of their own for any one article we could offer them. This, in my opinion, argues that they think themselves provided with all the necessaries of life, and that they have no superfluities."[2]

Indeed, the Aborigines paid little attention to the Cook expedition. Although Cook's ship, the *Endeavour*, was by far the largest and most complex human artifact ever seen on the east coast of Australia to that time, the natives displayed neither fear nor interest, and went on fishing or wandered into the forest. Some time after his first anchoring in Australia, Cook wrote,

> "We could know but very little of their customs, as we were never able to form any connections with them, they had not so much as touch'd the things we had left in their hutts. . . . All they seem'd to want was for us to be gone."[3]

Characteristically the Aborigines wore no clothing except for ornamental bands and belts made from hair or animal fur. In colder areas, however, animal skins were made into cloaks. Their weapons and tools varied from group to group (the boomerang, for example, might be an important weapon for some, a musical instrument for others, and unknown to others), but were invariably limited in number and variety. Basic equipment included spears, woomeras (spear-throwers), clubs, shields, stone axes, digging sticks, net bags, wooden bowls, and (in some areas) baskets. Desert dwellers had fewer possessions: a man might carry a woomera, a number of spears, and a throwing club; a woman would have a digging stick and a range of wooden containers. Often these items would be intricately inscribed with art forms of religious or magical significance.

Aborigines had no pottery. All cooking was done in ashes from wood fires. This type of society, basically lacking in tangible artifacts, creates great difficulties for archaeological recreation of its way of life. The essence of aboriginal society was in skills, rather than in hardware, with fire as a critical tool.

What the Aborigines lacked in material possessions (and their material culture surely ranked among the simplest in the world), they made up for in nonmaterial aspects. They participated in elaborate ceremonies of various kinds, and had an extraordinarily complex set of religious and magical beliefs, superstitions, and taboos, based largely upon an intricate oral mythology that emphasized the Dreamtime. The Dreamtime (or Dreaming) is a sophisticated and interconnected mosaic of knowledge, beliefs, and practices concerning the creativity of ancestral beings and the continuity and values of Aboriginal life. This rich cosmology provided the framework by which every

[2]Christopher Lloyd, ed., *The Voyages of Captain James Cook Round the World* (London: The Cresset Press, 1949), p. 87.

[3]Robert Hughes, *The Fatal Shore* (New York: Knopf, 1986), p. 54.

individual was bound, from before birth to after death, into an intimate, personal identification with the land and specific sites within it. The artistic expression of their spiritual beliefs involved cave paintings, wood carvings, unique bark and sand painting, rhythmic dances, and intricate charade-games and rituals.

The total population of Aborigines in Australia at the time of European contact is unknown; the most widely accepted estimate is 300,000, but some authorities place the number as high as 1.5 million. Their geographic distribution, density, and mobility were closely related to the availability of food, water, and other resources. Population density is estimated to have ranged from one person per square mile (2.6 km^2) in favored coastal areas to one person per 50 square miles (130 km^2) in the drier deserts.

Linguistic variety was great, with dozens of mutually unintelligible languages and dialects spoken in different areas. Despite the relatively small size of most tribelets and clans, and the almost complete lack of political organization, there would sometimes be gatherings of a thousand or more Aborigines for ceremonial or hunting purposes. Inter-group relations were usually peaceful, and recent evidence shows that there were some extensive trading networks, involving axes, boomerangs, shells, ochre, and nicotine-based plants.

It is tempting to visualize "noble savages" as living in tranquil harmony with their environment, but such a concept probably is no more valid in Australia than it is in other parts of the world. Consider the evidence: repeated deliberate burning significantly altered the floristic associations of the continent; the introduction of the dingo infused a supremely capable predator into a continental ecosystem that was replete with relatively vulnerable prey; the diprotodon (a giant marsupial) was exterminated; the thylacine ("Tasmanian wolf") was extirpated on the mainland.

Tranquil or not, after timeless eons of unchallenged hegemony, the Aborigines abruptly were confronted with alien invaders whose arrival signalled the conclusion of a way of life as surely as if extraterrestrial beings had landed on Earth. The most advanced nation in the world (the steam engine had just been invented in Britain) came face to face with a host of mini-tribes, none of which could even boil water.

The general result of European penetration and settlement on the aboriginal population was just what it has been all over the world where Europeans have brought the blessings of their civilization—the natives were depleted, debilitated, debauched, dispossessed, disregarded, disadvantaged, and discriminated against. The aboriginal culture and way of life was very different from that of the invaders, and vice versa, so that major misunderstandings were frequent. Although relations between the newcomers and the natives were mostly peaceful, direct conflicts did occur.[4] More

[4]In Tasmania, for example, it required less than half a century of white settlement to exterminate the people who had occupied the island for some 30,000 years; this was apparently the only true genocide in British colonial history.

lethal were the new diseases—whooping cough, measles, influenza, tuberculosis, and venereal diseases—which took a fearful toll. The death rate among Aborigines for the first century after contact was about 50% per generation. In addition, there was rapid erosion and destruction of the spiritually significant part of Aboriginal life, with consequent deterioration of tribal and individual vitality.

Discovery by Europeans

The recorded history of Australia is quiet and relatively uneventful. If one's concept of history is focussed on the dramatic, then Australia has no history. There were no invasions, no civil wars, no revolutions, not even a great deal of conflict between the indigenous inhabitants and colonizing Europeans. Australia is a nation that grew in relatively tame and orderly fashion from its first convict settlement to federation in little more than a century.

The ancient and remote Austral continent provided a secluded home for bizarre biota and hardy natives for dozens of centuries. This pristine land was touched by East Indian sailors and fishermen from time to time, but their imprint was almost nil. Despite speculation and prediction of the presence of a southern continent, European attention to the Southwest Pacific was slow in materializing; it was not until the 17th century that European navigators coasted Australian waters.

Legends and rumors concerning the existence of an Austral (southern) land mass had been prevalent sporadically since the 2nd century A.D. (Ptolemy's time). Ancient literature was replete with references to a mysterious southern continent called Terra Australis, although its location and dimensions were vague.

It is not known who was the first foreigner to set eyes on Australia, to "discover" the continent from the European point of view. As with various other parts of the Pacific, there is some speculation that Spanish and Portuguese navigators may have viewed sections of the Australian coastline before 1600. Indeed, it now seems likely that Portuguese sailors under Mendonca visited the east coast of Australia in 1522. However, no actual records of such occurrences have been found. The earliest evidence dates from 1606, when Luis Vaez de Torres sailed between Queensland and New Guinea in the strait that now bears his name, but even this contact was inadequately recorded, and the Torres Strait did not appear on maps until more than one and a half centuries later.

In that same year, a Dutch sea captain named Willem Jansz sailed down the west coast of Cape York Peninsula for some distance, and he is given credit by some authorities for "discovering" Australia. His opinion of the land was contemptuous ("no good to be done there") and did little to encourage future developments. During the succeeding three decades at least nine other Dutch vessels explored the Australian coast, most of them inadvertently, having been blown too far east when sailing with the prevailing westerlies from South Africa to Java. In this fashion the coastline came to be known and recorded incompletely, from the Great Bight around the western and northern sides to Cape York. The Dutch captains, however, were interested primarily in the East Indies, and they were generally unimpressed with the arid and semiarid coasts that they found. Still, they charted most of the west coast, and it appeared on Dutch maps by 1620.

Abel Tasman was the last of the Dutch sailors associated with the history of "New Holland," as Australia was then called. In the early 1640s his broad circumnavigation of the continent (he went around New Zealand and New Guinea at the same time) offered the first definite proof that it was not connected with any other major land mass. On the same voyage he discovered Tasmania, which he named Van Diemen's Land after the Dutch Governor of Java.

William Dampier was the first Englishman known to visit Australia. He explored along the northwest coast in 1688 and again in 1699. He experienced miserable weather and came away with little favorable to report to the Admiralty, discouraging further exploration for many years. Apart from these two contacts by Dampier, it was well over a century after Tasman's voyages before there was any other exploration of significance in Australian waters.

Captain James Cook, the greatest of the Pacific explorers, contributed significantly to the European settlement of Australia. He was anxious to prove the existence (or lack of same) of a great southern continent and its relationship with New Holland and New Zealand. In 1770, on the first of his three major voyages, he spent some five months along the east coast of Australia. He was the first European to spend much time on the humid margin of the continent and transmitted favorable reports to England. Cook claimed the land for George III, named it New South Wales, and charted much of the eastern coastline.

Initial Settlements

As a more or less direct result of Cook's reports, the colonization of Australia from Britain was begun. At the outset, the British government decided to people the Australian colonies for the most part with convicts. Thus the early immigrants were convicts who had been transported to Australia from the British isles for two reasons: to relieve overcrowding in British prisons and to supply cheap labor to the colonies. Thus most of the early settlers were not chosen for their skills, but rather to banish them from Britain.

The First Fleet consisted of an 11-ship prison convoy, which carried 736 convicts (548 male; 188 female) and 294 other persons (including marine guards) to a landing at Botany Bay in January 1788. Governor Arthur Phillip soon recognized the limitations of the area for settlement, and transferred the entire colony to the next bay northward (Port Jackson), where the Sydney settlement was founded and where the city of Sydney stands today (see Figure 3-1). This was the principal nucleus for the colonization of the continent.

The second settlement was established in the estuary of the Derwent River in Van Diemen's Land. Two groups of settlers, many of them convicts, were landed in 1803 and 1804, and Hobart Town was founded. This Tasmanian colony was designed as a convict settlement, but it was also proposed to test the possibilities of grain cultivation, timber export, and sealing. The final stimulus for settlement was fear that the French might try to settle there first.

Another colony was started in Van Diemen's Land in 1804, this time on the north coast. Some 180 persons, about half of them convicts, were landed. The settlement was shifted twice, but became stabilized in 1806 on the present site of Launceston. Other settlers soon came to the island, from the mainland of New South Wales, from Norfolk Island (where another convict colony had been established), and from Britain. By 1810 the population of Van Diemen's Land numbered more than 1,300, including about 250 convicts.

Figure 3-1 Exploration and Settlement

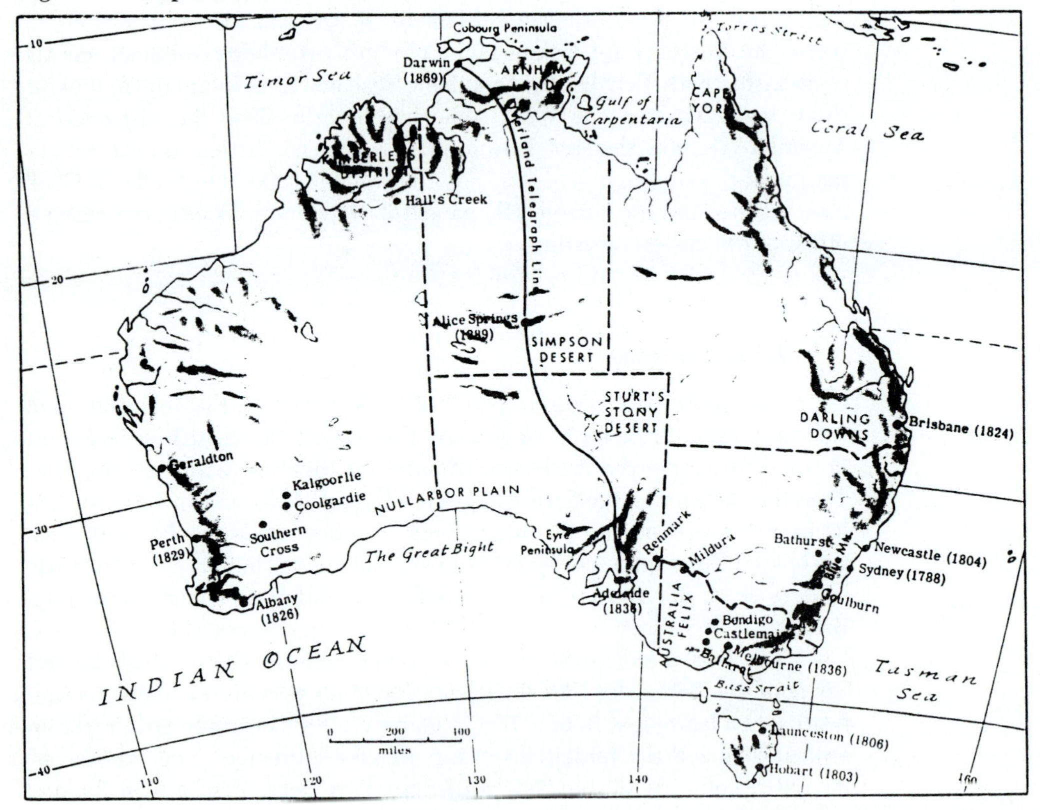

The second settlement on the mainland was originally more or less an "out-station" from Sydney. After a couple of false starts, a small number of convicts was shifted to Newcastle, at the mouth of the Hunter River about 100 miles (160 km) north of Sydney. An abundance of nearby bituminous coal helped to assure the success of the settlement, even though it was primarily a penal station for a number of years.

Several attempts were made to settle in what is now Victoria (it was called the Port Phillip District of New South Wales until separation in 1851), from as early as 1803. The first involved more than 400 people, but the colony was abandoned within four months and the settlers moved on to Hobart. Several other abortive settlements were made before the first "permanent" squatters (settlers without legal land grants or even governmental permission) arrived in the early 1830s. Many more squatters, often from Van Diemen's Land, arrived in the mid-1830s; in 1836, at which time the district had some 200 European inhabitants, an official government settlement was proclaimed. The settlement nucleus, originally called Bearbrass, was renamed Melbourne in 1837 and grew more rapidly than any of the other original Australian settlement centers.

The first settlement in what is now Queensland dates from 1824, when 30 convicts and their guards were landed at Redcliffe, on the southeastern coast. Later that year Governor Brisbane (of New South Wales) ordered the removal of the colony, which was subsequently named after him, to a more propitious location a few miles inland. By 1830 the Brisbane colony contained about 1,000 convicts and 100 soldiers. The first free settlers began to "squat" in the Darling Downs (inland from Brisbane) in 1838; four years later the prison colony was abandoned and the area was officially thrown open to free settlement.

Several settlements were attempted on the north central coast of Australia, to the north and east of the present location of Darwin, in the 1820s and 1830s. Harsh conditions, remote location, and human foibles resulted in their all being abandoned before the midpoint of the 19th century.

In order to enhance Britain's claim to the entire continent and to forestall anticipated French settlement, a group of settlers was landed at Frederickstown (the present-day Albany), in what is now Western Australia, in 1826. A more permanent settlement was established on the Swan River, where Perth is now located, three years later. The colony did not prosper, however. There were notable plans for free settlement, but a combination of ignorance, lack of manpower, and minimal government support diluted the effort. The population had reached 4,000 in 1830, but dwindled to less than 1,000 by 1832.

The last of the initial settlements was in many ways the most interesting and innovative. South Australia was founded in 1836 as a planned colony based on free enterprise and an orderly design for settlement. It was unique among the initial colonies in that convicts were never involved. Settlement was organized by a land settlement company, conceived by British social theoreticians (especially Edward Gibbon Wakefield), and backed by British capitalists. The harsh realities of the new land presented many difficulties, but the colony struggled forward until 1842, when an administrative reorganization centralized more authority under the Crown. The colony registered continued, if erratic, growth in its early years.

Exploration by Sea and by Land

The coasts of Australia, understandably, were wellknown long before the early settlers had any real concept of the inland. Most of the coastline was charted, at least in a generalized fashion, before the end of the 18th century. The south coast was the last to be surveyed. George Bass and Matthew Flinders sailed through Bass Strait for the first time in 1798, and proved that Van Diemen's Land was an island. Flinders surveyed most of the rest of the south coast in 1800–1802. The French navigator, Baudin, also charted much of that coast in 1802. Ironically, the very last part of the Australian coast to be surveyed was around the mouth of the continent's only large river (the Murray), at Encounter Bay (where Flinders and Baudin met).

Exploration by land was incidental to settlement and began only after the Sydney penal colony was thoroughly established. Mostly, the overland exploration was practical in intent and empirical in method. The initial goal was to find good farming land, and later it was to discover good pastoral land. When explorers were successful in these endeavors, settlers were rarely far behind.

Overland exploration was beset by numerous difficulties. The most immediate problem was to break out of the relatively small lowland that encircled Port Jackson on the three landward sides. Rugged, sterile sandstone ranges cut the Sydney colony off from the interior. The tumbled gorges and scarps of the Blue Mountains presented precipitous slopes that defied penetration for more than a quarter of a century after initial settlement. Although the mountains were finally crossed for the first time in 1813, they remained a formidable obstacle to effective penetration of the interior for some time to come.

Once these mountains were breached, the true nature of the Australian continent began to assert itself. Further penetration was slow because of a scarcity of that most precious of commodities—water. Living off the land was often difficult because food resources were inadequate. Other problems included high temperatures, persistently bothersome insects, and, sometimes, Aboriginal hostility.

The first of the major inland explorers was Charles Sturt, who spent the better part of two decades (the 1820s and 1830s) investigating the interior of New South Wales. His expeditions contributed immensely to an understanding of the complex river pattern and regime in that region. He finally succeeded in tracing the River Murray to its mouth, perhaps the most important single exploration in Australian history.

At about this same time there were various less ambitious explorations from the Sydney colony southward into the Port Phillip District, westward to the plains of Goulburn and Bathurst, and northward toward Queensland. Reports from these expeditions stimulated the movement of settlers in all three directions; although initial response often was only a trickle of migrants, these streams became floods before long.

Most settlement stayed fairly close to the east coast then, but explorers became more daring, and began to make deeper penetrations into the Outback. There was a continuing interest in the inland. Although from the fringe it looked remarkably like a semi-desert, hope sprung eternally that it would prove to contain fertile farm or grazing land. Many people grappled with personal obsessions that exploration of the interior would reveal an inland sea, or an inland river system, or a high mountain range serving as a watershed divide. All of these hopes were forlorn, of course, but they nourished fruitful results, for they stimulated the exploration of the continent.

The first major exploration northward was that of Ludwig Leichhardt, a German scientist of mystical bent, who traversed from the Darling Downs of southeastern Queensland to the Cobourg Peninsula of Arnhem Land in 1844–1845. His route was paralleled in part by an 1845–1846 expedition led by Major Thomas Mitchell, which succeeded in exploring a large part of interior Queensland. This was only one of a number of significant explorations by Mitchell, who had previously surveyed much of the interior of New South Wales and had penetrated into the fertile western portion of Victoria, which he called Australia Felix.

The first east-west crossing of the continent was accomplished by E. J. Eyre in the early 1840s. He was the leader of a South Australian government expedition that had been intended to seek a feasible livestock route to the west coast from Adelaide. At Eyre's insistence, the purpose of the expedition was changed to a search for the center of the continent. However, unusual rains bogged the party at Lake Torrens, so they had to go westward after all. Eyre finally succeeded in crossing the Nullarbor Plain, accompanied by a single Aborigine, and he reached Albany after one of the classic journeys in the history of exploration. Unfortunately, this remarkable walk resulted in little information of practical value, as the participants were forced to follow the coastline most of the way.

Charles Sturt had shifted his interest from New South Wales to South Australia by the 1840s. His last expedition, an heroic failure to reach the center of Australia, covered 18 months in 1844–1846. He was stopped by drought conditions and turned back by the gibbered surface of Sturt's Stony Desert and the sand ridges of the Simpson Desert.

Ludwig Leichhardt's final imprint on the Australian continent was a question mark. In 1848 he led an expedition westward from Brisbane, in an attempt to cross to the west coast. With six companions and nearly 300 head of livestock, he disappeared and was never heard from again. No definite traces of the expedition were ever found, and its actual fate is the great mystery of Australian exploration history.

The first successful south-north crossing of the continent was accomplished by the famous Burke and Wills expedition in 1860–1861. After starting from Melbourne with an unwieldy expedition of 20 men, 26 camels, and 23 horses, they slimmed down to a smaller party (four men, six camels, and one horse) at Cooper's Creek and made a "dash" to the north coast. Favored by good weather, they reached the Gulf of Carpentaria without major difficulty, but the return journey was an almost complete tragedy. The horse, four of the camels, and three of the men perished. Furthermore, their surveying had been inadequate and the leader (Robert Burke) had not even kept a journal.

Simultaneously, an expedition led by John McDouall Stuart was working its way northward from Adelaide. This was Stuart's sixth penetration of the interior (he had reached the center of the continent on his fourth journey in 1860), and he successfully reached the north coast in 1862. His careful survey and favorable report led directly to settlement in the Northern Territory and to the construction of the Overland Telegraph Line, completed in 1872, from Adelaide to Darwin. This line, in a sense, separated the semi-explored eastern half of the continent from the totally unexplored western half. To that date only Eyre had crossed Australia from east to west, and he had been in sight of the southern coastline essentially all the way.

Colonel P.E. Warburton led a significant east-west crossing in 1873–1874. His expedition traveled from Adelaide to Alice Springs and thence westward. Warburton had planned to reach the central coast of Western Australia, but dry conditions kept forcing the party northwestward. It was a fearful journey, and although none of the seven men perished, all just barely survived. This was the first Australian exploring expedition to use camels as the sole means of transport, and it is generally agreed that the hardy nature of these beasts was the only factor that prevented the expedition from ending in total tragedy.

John Forrest, a young and vigorous Western Australian, reversed the normal route and led two expeditions from west to east in the early 1870s. He traveled from Perth to Adelaide in 1872 and from Geraldton to the Overland Telegraph Line two years later. Forrest was later appointed the first premier of the colony, and after federation he served in the federal parliament for eighteen years.

Ernest Giles, a man of great drive, led five separate expeditions in South Australia and Western Australia in the 1870s, including a double crossing of the continent (east-west and west-east). His explorations led to a filling in of the last major blanks on the map of Australia, and he is therefore generally regarded as the last of the major Australian explorers. Thus ended the heroic era.

The Early Spread of Settlement

The settlement at Sydney was established as a penal colony, but almost immediately it began to be something more. The convicts were not able to make themselves self-sufficient, a fact which hastened the rather stormy transition from jail to free colony. Agricultural production was the greatest need, and there were many attempts at farming. However, for the first 30 years or so the settlement was restricted to a small area around the original nucleus, which contained an inadequate amount of good farmland. There were limited tracts of shale and silt that harbored soils that would support crops, but the settlement was essentially hemmed in by a vast belt of unproductive sandstones and abrupt slopes. Thus foodstuffs and other essential supplies had to be brought from the other side of the world for several decades, a necessity which made life in the new land precarious.

Very early it became clear that the area would never be fruitful for farming, so emphasis turned to stock raising. Several varieties of livestock were imported, but the Merino sheep turned out to be far and away the most suitable. The emphasis on breeding and raising sheep for wool production, particularly fostered by Elizabeth Macarthur and her husband, John (an ex-Army officer and large landowner), in the early years of the 19th century, changed the entire course of development of the colony, and sheep raising became the economic lifeblood of the continent for many decades to come.

As sheep raising developed, there was much desire to push beyond the narrow confines of the Sydney area. Official governmental policy was against this tendency, because of difficulties of managing and policing away from the centralized settlement. However, during the second decade of the 19th century, explorers were finally able to open routes across the barrier of the Blue Mountains. Settlement quickly spread to the Bathurst Plains on the west and to the Goulburn Plains on the southwest, and within a few years the frontier of expansion was advancing outward in all directions from Sydney. By the 1830s, there were more free settlers than convicts in Australia (see Figure 3-2), and by 1840 pastoralists were ensconced as far south as Australia Felix (the name given by the explorer Thomas Mitchell to a large tract of excellent pastoral and farming land in what is now western Victoria), as far west as Spencer Gulf, and as far north as the Darling Downs. Thus the central section of New South Wales, as the entire eastern part of the continent was then called, experienced much settlement expansion during the 1830s and 1840s. All of the humid and subhumid portions of the region were occupied, and even the semi-arid plains of the west received a share of the expanding population. Some of this activity involved large land settlement companies from Britain, but most was of the individual settler and squatter type. Presaging even greater mining activity in all the Australian colonies a few years later, coal mining got an early start in the Hunter Valley. The first boatload of coal was shipped to Sydney in 1801, and by 1814 there was a thriving coal trade with India. The Hunter Valley was very attractive to agricultural settlers from the 1820s onward.

Van Diemen's Land, although primarily a penal colony, was producing both sheep and wheat at an early stage. Settlement spread up the Derwent and Macquarie valleys and along the north coast. Whaling and sealing thrived at various coastal towns, and wheat was being exported to the mainland by

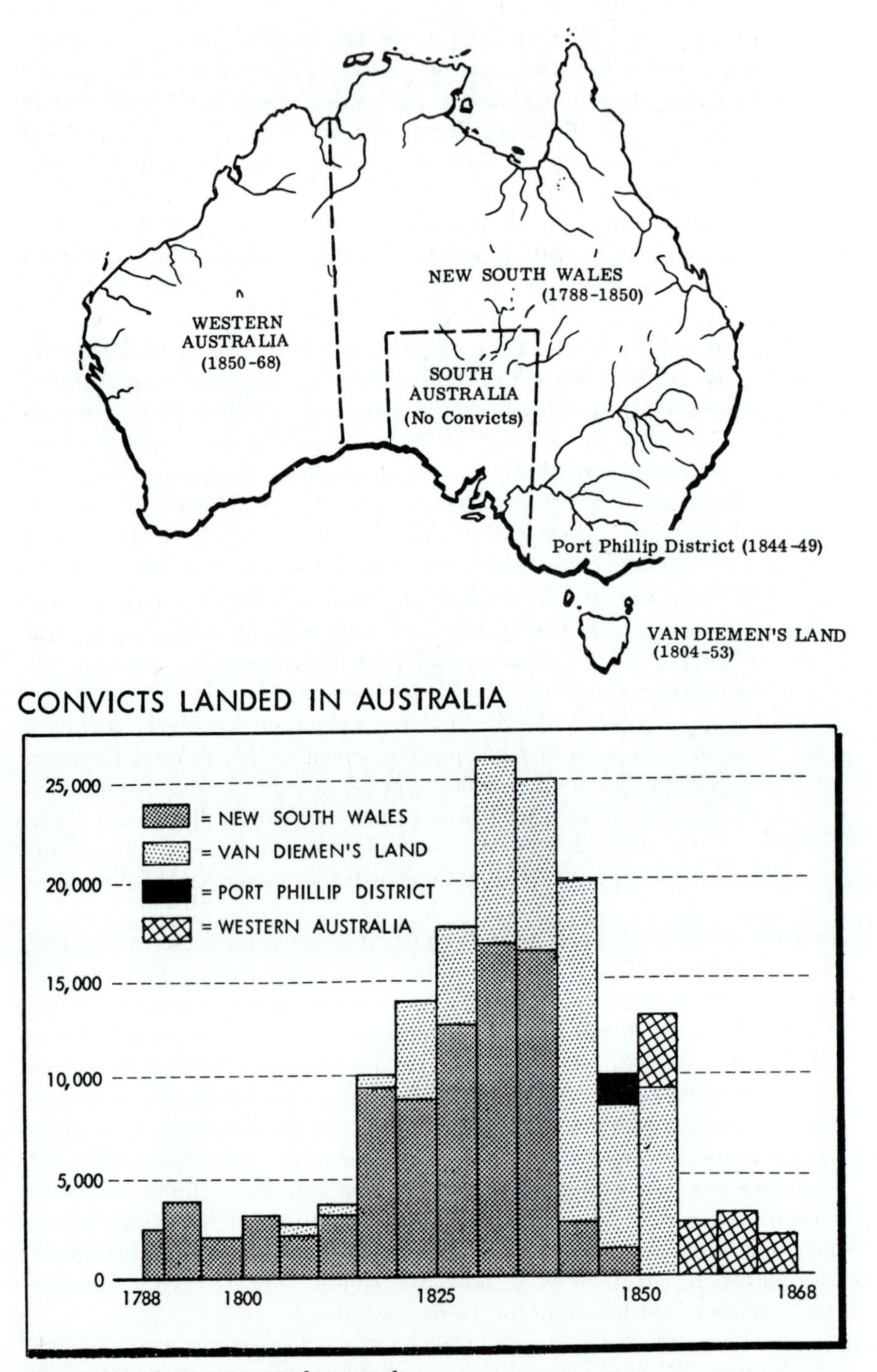

Figure 3-2 Convicts in Early Australia

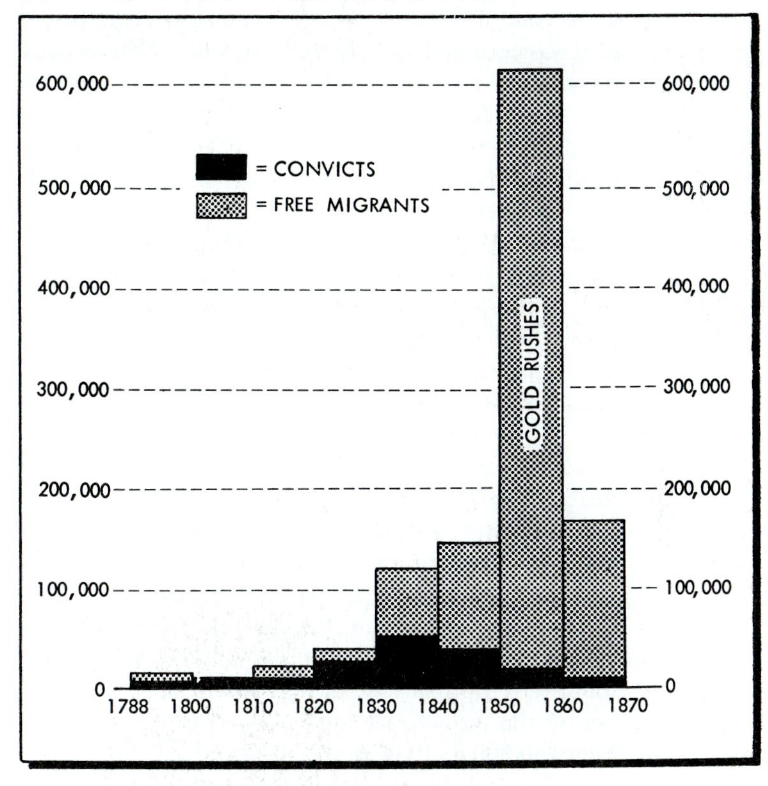

Adapted from Ian Wynd and Joyce Wood, *A Map History of Australia*, 2nd edition. (Melbourne: Oxford University Press, 1967), P. 11. By permission of the publisher.

Figure 3-2 (*cont.*)

1820. Agitation for separation from New South Wales grew, and Van Diemen's Land (the name was not changed to Tasmania for another 30 years) became a separate colony in 1825, at which time there were some 15,000 Europeans on the island, nearly half of them still convicts. Fights between settlers and Aborigines reached a peak about this time, and there were other civil problems; but population, livestock numbers, and crop production continued to expand with considerable rapidity.

Although a few squatters had come to the Port Phillip District early, the population was still very sparse around Melbourne until the latter part of the 1830s. The number of Europeans in the district increased from about 200 in 1836 to nearly 10,000 in 1840; sheep numbers in this same period grew from 25,000 to 800,000. Mitchell's favorable report on Australia Felix was a major stimulus to settlement expansion. Here, as always, exploration was largely at the behest of potential settlers, who followed close on the heels of the explorers. The tide of squatters and other settlers spread rapidly in the central and western parts of the district, and most of the usable pastoral land was occupied by 1845. In 1851, when Victoria became a separate colony, it contained an estimated 77,000 settlers and 5,000,000 sheep.

From the time the Moreton Bay area was opened to freemen in the early 1840s, the settlement history of Queensland was one of cattle expansion to the north, northwest, and west. Overlanders and their herds followed such explorers as Leichhardt and Mitchell, Kennedy and Gregory, Landsborough and McKinlay, and took up large pastoral holdings with or without proper title to the land. Indeed, the matter of inadequate land titles for white settlers was a contentious issue for many years, but it is now overshadowed by legal questions concerning Aboriginal land title (due to the almost total lack of treaties with the indigenous population), which has emerged in the 1990s as one of the most controversial problems facing the Australian nation.

A scarcity of labor and inadequate contact with Sydney led to agitation for separation from New South Wales. Queensland became a separate colony in 1859, with a European population of about 25,000.

In South Australia, as we have seen, the first years were difficult ones. Settlement expansion was carefully controlled so that there was orderly growth rather than "leapfrogging" outward from Adelaide. However, economic improvement was slow until the mid-1840s, when sheep, wheat, and copper contributed to prosperity and to settlement expansion eastward and southward, but particularly northward. Between 1840 and 1850 the population of the colony increased from 14,000 to 63,000, the number of sheep from 200,000 to 1,000,000, and the wheat acreage from 1,000 to 41,000.

The colony of Western Australia experienced the most desultory growth pattern of all. Unproductive soil and dense stands of timber inhibited much settlement expansion, and the colony all but stagnated during the 1830s and 1840s. Eventually convicts were brought in in order to provide cheap labor for economic stimulation, so Western Australia became a penal colony between 1850 and 1868; still, settlement expansion was limited until the gold rush days later in the century.

Eastern Gold Rushes

Although Australia's settlement history was relatively peaceful, it was not without excitement, and the most exciting and dynamic times were associated with the discovery and exploitation of gold deposits. Each of the major gold rushes followed a similar pattern—hopeful men stampeded to the diggings from near and far, coming from all over Australia, from California, from Britain, and from China. The population expanded spectacularly for a few years, and then the counterflow began. Most of the miners either drifted away or settled down into more mundane occupations such as farmer, pastoralist, or storekeeper. The aftermath of a gold rush, then, invariably made a significant contribution to settlement expansion, to increased agricultural and pastoral production, and to urban growth.

The first and greatest sequence of gold rushes took place in Victoria, beginning in 1851. There were numerous strikes, but the major ones were at Ballarat, Bendigo, and Castlemaine. The gold was mostly in alluvial deposits, so miners needed only simple equipment. Literally hundreds of thousands of people were attracted to the colony—from Tasmania, from South Australia, and from all over the world—including some 20,000 Chinese. Indeed, by 1861 there were 40,000 Chinese in the Australian colonies, giving them third place (after British and Germans) among all non-Aboriginal ethnic groups on the continent. As almost all the Chinese were male, it can be calculated that more than 10% of all non-Aboriginal males in Australia were Chinese at this time. It was not long before all of the eastern colonies had enacted Oriental exclusion immigration restrictions.

The population of Victoria increased from 70,000 to half a million in less than eight years. The peak of gold production was achieved in 1856, but the population influx gave a matchless impetus to the agricultural and industrial sectors of the economy.

New South Wales experienced gold rushes at the same time as Victoria, in the early 1850s. The strikes were smaller, more scattered, and shorter lived, with peak production dating from 1852. Nevertheless, significant population growth was recorded, and settlement expansion was considerably accelerated.

Gold was discovered in each of the other eastern colonies during this same decade, but the yields were small and the rushes were insignificant in comparison to those of Victoria and New South Wales.

Filling in the Pattern

In New South Wales, Victoria, and Tasmania, and to a lesser extent in the three other colonies, the middle of the 19th century was a time of considerable unrest and disorder, analogous in many ways to the "Wild West" period of American history. Land ownership patterns were at the root of the problem. Government-sponsored land settlement schemes allowed some pastoralists to amass huge holdings. Other settlers often "squatted" on the land illegally, and the entire matter of land alienation and ownership was strife-laden. There was much agitation for small land grants to be made to landless individuals, so that they could make a start as farmers or pastoralists.

This general situation led to a certain amount of land reform, and it also resulted in the "bushranger" era of Australian history. Bushrangers were essentially outlaws or highwaymen who robbed banks, towns, and especially coaches. Although part of the rural scene in the eastern colonies from the earliest days of settlement (the first bushrangers were escaped convicts), it wasn't until the 1860s that bushranging reached its heyday of activity. Many bushrangers of that period (such as Ned Kelly, Ben Hall, and Captain Moonlight) have become leading Australian folk heroes.

During the latter half of the 19th century the settlement of Queensland was diffused and intensified by several factors. Cattlemen continued to settle all over the colony, pushing the frontier outward and filling in many of the gaps. Furthermore, sheep were brought into the central plains in increasing numbers, and the foundations for the present distribution pattern (sheep in the center and south; cattle in the west, north, and east) were established. Specialized tropical agriculture got its start around Brisbane in 1862 with the first sugar cane plantation. Despite a shortage of local labor, cane production spread northward to occupy most of the fertile coastal valleys between the New South Wales border and Cape York Peninsula. Natives of various Melanesian islands, called "kanakas," were imported to work on the sugar plantations, causing considerable uproar among the labor-conscious European population. After a quarter of a century kanakas were prohibited from entry, but nearly 50,000 of them had been brought in by that time. Despite Queensland's vast extent, most of the colony except for Cape York was occupied by settlers by the 1880s.

The spread of settlement in South Australia became less orderly with the passage of time. The Eyre Peninsula attracted sheepmen and farmers; cattle raising became important in the cooler, more humid areas of the southeast; the lower Murray basin attracted settlers and entrepreneurs, partly for farming and partly to participate in river commerce; and sheep-wheat farmers spread northward into the Flinders Ranges. At its greatest extent, wheat was being grown as far as Wilpena Pound, but the unusually ample rainfall that had attracted farmers that far north soon gave way to more usual conditions, resulting in a rollback of the farming frontier to the south of Goyder's Line (a line marking the northward margin of country deemed by the colony's Surveyor-General to be capable of growing wheat in normal years). A significant drain on the colony's budget during these years was the administration of the Northern Territory, which was governed by South Australia from 1862 till 1911.

Successful settlement was tardy in the Northern Territory. Even the completion of the Overland Telegraph Line did not serve as much of a stimulus. The town of Palmerston (later renamed Darwin) was established in 1869, and Stuart (later to become Alice Springs) was not laid out until two decades later. There were various gold discoveries in the 1860s and 1870s, but they were not nearly as notable as the strikes in the eastern colonies. The gold did attract Chinese, however, and there was a considerable influx of Orientals in the last quarter of the nineteenth century. Pastoralists did not pay much attention to the Northern Territory until 1872, when the first cattle stations were established in the Alice Springs district (the "Centre"). In the 1880s, herds of cattle began to be overlanded from Queensland. In a relatively short time much of the northern part of the territory (the "Top End") was divided into pastoral properties, as was much of the Centre. The territory's first railway was a 100-mile (160-km) line built from Palmerston to Pine Creek in 1889.

Gold was the most important stimulus to settlement in Western Australia. Although Perth had been connected to Albany by a 250-mile (400-km) road through the impressive jarrah and karri forests since the early years of the colony, the expansion of settlement had been slow. Pastoral properties were sporadically occupied in the hinterland, but rapid development did not occur till late in the century. The first of the Western Australian gold strikes took place in the Kimberleys district of the far north, at Hall's Creek in 1885. It was short-lived but was followed by a number of other discoveries further south, the most important of the early finds being the Southern Cross field 200 miles (320 km) east of Perth. Following the geologic trend eastward, the rich alluvial gold of Coolgardie was discovered in 1892, and in the following year was found the richest of them all, the Golden Mile of Kalgoorlie.

Prospectors and miners swarmed to these goldfields, as they had to those of Victoria and New South Wales. The population of Western Australia increased by 35,000 in 1896 alone. The improvement of the roadway and the building of a railway from Perth, along with the construction of a water supply pipeline from the Darling Scarp area to Kalgoorlie, quickened the pace of economic activity and made it possible to open up a vast area of farming and pastoral land between the coast and the goldfields. Settlers continued to push the farming frontier north and south from the railway line, and Western Australia's "wheat belt" developed rapidly. There were a number of other gold strikes in the central part of the colony, especially in the Murchison, Ashburton, and Pilbara districts; all were puny in comparison to the earlier discoveries, but each of the new fields stimulated settlement expansion. While pastoralists and farmers were spreading over the southern and central parts of the colony, pastoral occupance of the north (Kimberleys district) was also beginning. Cattle were overlanded to the Kimberleys for the first time in the 1880s and, despite several fierce setbacks due to such factors as cattle ticks and red-water fever, most of the grazing land was occupied by the turn of the century.

Railroads made an important contribution to both economic development and settlement expansion. The first short lines were constructed in the 1850s, and each colony continued a fairly rapid pace of railway building until there were over 10,000 miles (16,000 km) of track in Australia as the 20th century dawned. Unfortunately, the lines of the various colonies were unintegrated, with the result that different track widths were chosen, and the problem of non-uniform gauge has plagued Australia to the present day. Most of the major rail lines were designed to serve as gathering systems to funnel rural produce to the capital city. Thus they fostered the centralized growth of a single city in each colony, another pattern that has persisted.

Irrigation agriculture provided a further stimulus to settlement expansion. The first major schemes were developed in the Renmark-Mildura area of the middle Murray valley by the Chaffey Brothers beginning in the late

1880s. Most of the other irrigated farming areas in Australia are 20th century developments. The largest scheme was in the middle section of the Murrumbidgee Valley (called the MIA) in New South Wales, but there were numerous smaller developments, especially in Victoria and Queensland.

The six Australian colonies finally agreed to unite, and federation was achieved in 1901. At that time the total population of the new country was less than 2,000,000. The settlement pattern, however, was well developed; 20th century additions have been mostly in the form of intensification rather than expansion into new areas.

Population Growth in the 19th Century

The period of Australian history prior to the 1820s is generally referred to as the "transportation" era because it was predominantly a time of convict importation, with little governmental encouragement of free settlement. By 1821, the European population of New South Wales (which still included Van Diemen's Land and the Port Phillip District) was about 36,000; of these settlers, some 21,000 were convicts, 7,000 were emancipists (prisoners who had served their sentences and been freed), 1,500 were free settlers, and the remainder were children.

Transportation of convicts continued until 1868, amounting to a cumulative total of more than 160,000 prisoners during the 80 years of the practice, but the immigration of free settlers became increasingly important after 1821. Immigration totals fluctuated through the years, but the steady increase in total population is indicative of its continuance. Almost all of the migrants to Australia during the first half century were from the British Isles; indeed, New South Wales received essentially no immigrants who were not either British or Irish during this time. In South Australia, the approach was different; the government encouraged Continental migrants, especially Germans, almost from the outset. By 1850 more than 4,000 Germans had come to South Australia, although these amounted to less than 10% of the colony's population. Germans also began to immigrate to the Port Phillip District in the late 1840s.

During the latter half of the 19th century, the population of Australia increased from 400,000 to 3,700,000; natural increase (excess of births over deaths) amounted to less than 60% of the increment, whereas net immigration (excess of immigration over emigration) accounted for more than 40% of the growth. During this time of significant total expansion, there were also major changes in the composition of the population. The British or Irish origin of most of the immigrants continued to be dominant, but its proportion declined, and several other sources became important. Germans comprised the largest non-British element, settling in all the colonies in considerable numbers. Chinese also were notable at this time; they entered the eastern colonies in great numbers to participate in the gold rushes, and Chinese-born immigrants actually outnumbered German-born immigrants from the 1850s until the 1880s. In the 1870s and 1880s a more diverse pattern of

European minorities began to arrive in Australia as a result of Scandinavian and Italian migration. In addition, several tens of thousands of kanakas were brought into Queensland to work in the cane fields during this period.

By the turn of the century, the nonindigenous population of the country was 75% Australian-born, ranging from about 80% in Tasmania to about 70% in Queensland. Another 20% of the total had been born in the United Kingdom; mostly in England, many in Ireland, some in Scotland, and a few in Wales. The non-British 5% of the population was proportionately declining at this time; Chinese were decreasing rapidly, Germans and Scandinavians more slowly, and Italians were continuing to increase, heralding the expanded immigration from southern Europe that marked the 20th century.

Population Growth in the 20th Century

With the beginning of the 20th century came federation of the colonies and a new outlook for the country. Net immigration, however, which had been on a downward trend for two decades, reached a negative value (i.e., a net emigration figure) for the first time in history, in part reflecting repatriation of kanakas and emigration of Chinese. The new states once again emphasized British immigration, and each of them developed policies for assisting British migrants. The migrant flow soon began to increase rapidly once again, and reached a total in excess of 130,000 per year just before the outbreak of World War I.

The war effectively stopped all immigration to Australia for half a decade, but the post-war period was one of migration acceleration. Net immigration totals increased to an annual average of about 40,000 during the 1920s. The migrants were primarily British; in most years fewer than 15% were from continental Europe, largely Germany, Scandinavia, and Italy.

The depression years of the 1930s were marked by reduced migration movements, and early in the decade there were several years of net emigration from Australia. After 1936 immigration increased, and for the first time migrants from continental Europe outnumbered those from the British Isles. Once again, however, the flow pattern was interrupted by a world war, and migration was at a virtual standstill from 1939 till 1945.

Since the war there has been a continuing series of assisted migration schemes worked out between Australian state and federal governments on the one hand and various European countries on the other to foster and subsidize immigration. The result has been an almost continually rising immigration curve, although it should be pointed out that the emigration curve has followed the same trend at a lower level.

The Contemporary Population of Australia

From a pre-contact total of perhaps 300,000 Aborigines and no Europeans, the population of Australia has grown in two centuries to more than

17,000,000, less than 2% of whom are Aboriginal. The present population mix, then, is essentially the result of long-continued immigration, the pattern of which has fluctuated significantly through the years.

The major demographic characteristics of Australia are generally similar to those of other industrialized ("developed") nations, with a few notable exceptions. It is striking that such a large country (sixth largest in the world) should have such a small population (43rd largest in the world), although the environmental and historical reasons for this anomaly are clear. The overall population density is about four persons per square mile, easily the least of any significant nation.

The general pattern of population distribution strongly reflects environmental (particularly climatic) influences (see Figure 3-3). The drier parts of

Figure 3-3 Population Density

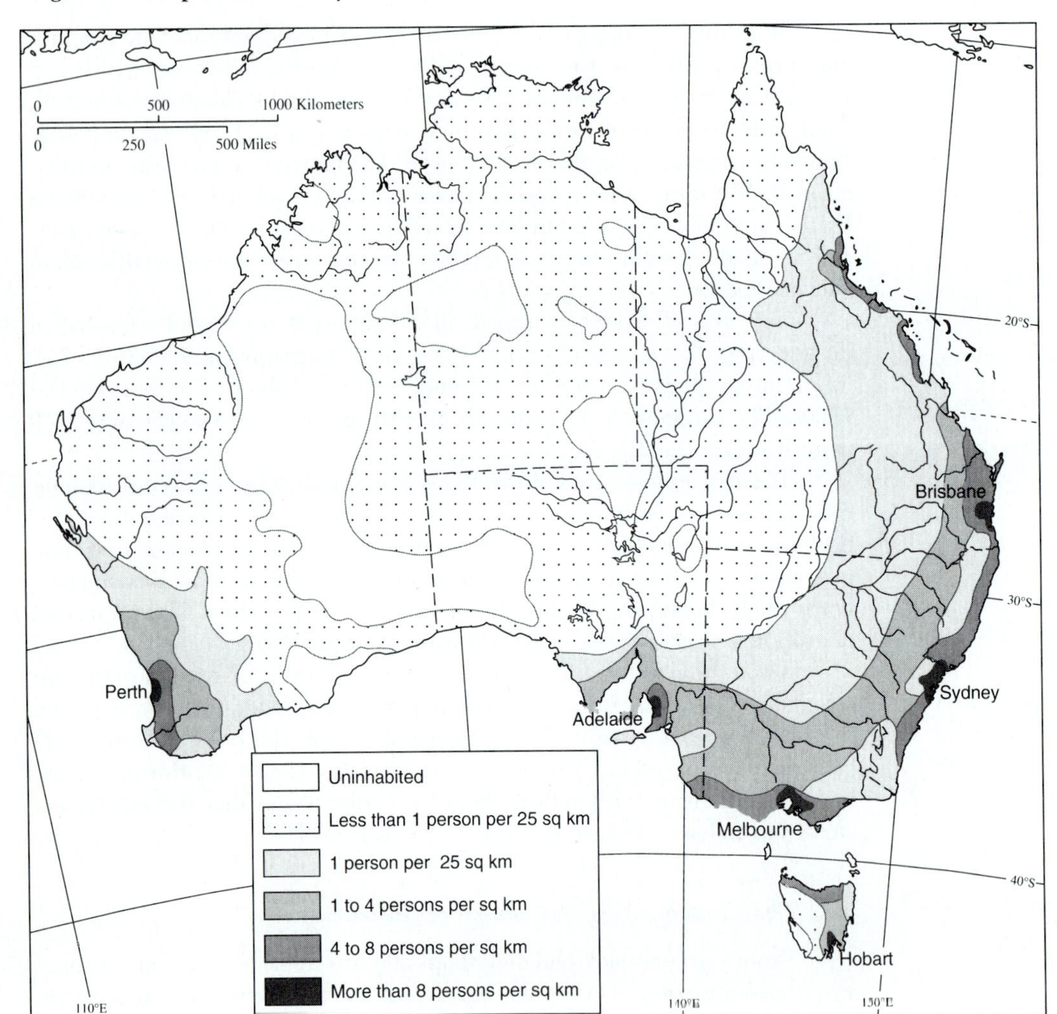

the continent are less productive than the wetter parts for the primary industries (farming and livestock raising), and therefore are sparsely populated. The humid areas, particularly the mid-latitude portions, permit more fruitful use of the land and are occupied in moderate density. The only areas of high-density settlement are urban or urban-related.

Thus the generalized population distribution pattern shows an arc of moderate density extending around the southeastern coast from Rockhampton (Queensland) to Whyalla (South Australia), with the principal concentrations between Newcastle (New South Wales) and Geelong (Victoria). The major urban centers punctuate this crescent as nodes of high density, but there are also areas of sparseness, such as the Australian Alps in New South Wales and Victoria. Beyond the arc, moderate population density appears in the southeastern and northern coastal portions of Tasmania, in the southwest of Western Australia, and in various east coast valleys of Queensland. Most of the remainder of the continent, some 80% of its area, is very thinly peopled.

For some years after World War II the annual growth rate of the Australian population was between 2% and 2.5%, but more recently it has slowed down to annual rate of about 1.25% (see Figure 3-4), which is consonant with other "developed" countries outside Europe (e.g., New Zealand, Canada, and the United States). Natural increase has been the larger component of population growth, amounting to about two-thirds of the total in recent years. In common with most industrialized countries, the Australian rate of natural increase has been diminishing as the birth rate declined; the natural increase rate is now about 8 per 1,000 persons per year. This reflects a balance between a birth rate of about 16 per 1,000 per year and a death rate of some 8 per 1,000 per year.

In terms of immigration, Australia has been one of the leading recipient countries of the world throughout most of its recorded history. It has been

Figure 3-4 Population Through the Years

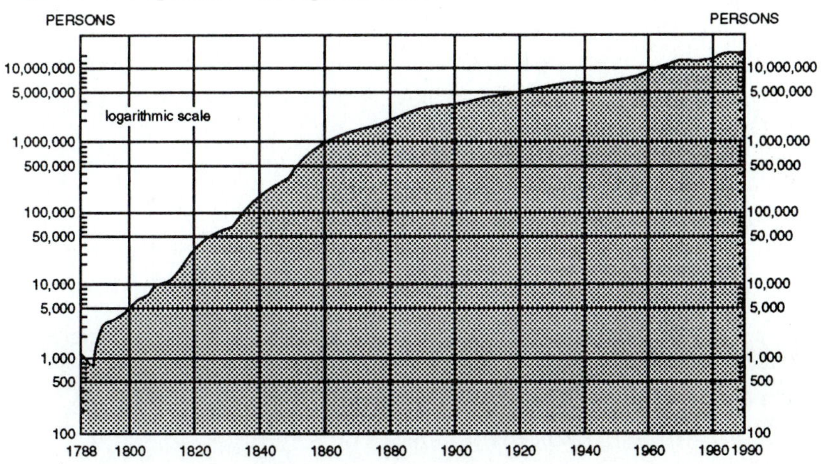

a prominent goal of European migrants for many decades, and now is attracting an increasing flow from Asia. In recent years, net immigration has accounted for about one-third of the population growth rate. More than one-fifth of the present population is not Australian-born. The net increase from immigration since 1975 has been about 90,000 per year, but in the late 1980s the government expanded immigration intake to nearly 150,000 annually. This gave Australia the distinction of having the highest per capita immigration quota in the world: 0.8% of its population annually (this compares with about 0.3% for the United States, for example), although this quota was reduced in the early 1990s to less than 100,000 per year.

The majority of all immigrants came from the British Isles until after World War II, when the balance shifted to continental Europe. The largest number of non-British migrants since the war have been Italians, followed by Greeks, Dutch, Yugoslavs, Germans, and Poles. More recently, there has been a significant upturn in immigration from Asia, particularly Vietnam, Malaysia, and Hong Kong. By 1990, more than one-third of all immigrants were from Asia.

The current population of Australia, then, is ethnically varied, but the ethnicity is largely of European origin. British stock is still dominant, comprising well over one-third of the total population. Continental Europeans are broadly represented, with only a few exceptions, most notably people of French origin. There is a small, but rapidly expanding, component of Asians. The Aboriginal component of the population is also growing rapidly.

Perhaps the most striking feature of Australian demography is the centralization of its population; some 62% of the total population resides in the seven capital cities. Another 24% lives in smaller urban areas, and only 14% can be classed as rural. Such metropolitan centralization has been a long-established characteristic, as an historic holdover from the days of separate colonies. Only during the 1980s, for the first time, did a slight counter-trend set in.

Focus Box: Contemporary Aboriginal Issues

A significant minority of the sparse population of the Australian inland consists of Aborigines. We noted earlier that the original indigenous population was severely decimated through the years, reaching a low point of about 40,000 at the beginning of the present century. Since that time, the trend has been reversed, and the Aboriginal population has been increasing at a much faster rate than that of the nation as a whole. The total number of Aborigines is unknown because they have been severely under-enumerated in the censuses.[5] The 1991 census indicated a national total of about 200,000 Aborigines and Torres Strait Islanders (inhabitants of a number of small islands in the Torres Strait between Cape York and New Guinea; there are about 25,000 Torres Strait Islanders). However, it is likely that the actual total is closer to 400,000.

It is probable that no Aborigines still lead a completely nomadic hunting-and-gathering existence, although a few thousand inhabit the larger and remoter Outback reserves, maintaining tribal identity and having minimal contact with non-Aborigines.[6] Several thousands live on more settled and less remote reserves, often fully supported by the government and slowly learning modern skills. Many thousands are dwellers on the fringe of civilization, loosely attached to towns or cattle stations or missions. Many more thousands have moved to the cities, where they mostly consist of a semi-invisible, poverty-ridden subculture. A relatively small number of Aborigines have been assimilated into the general Australian urban society.

Another notable feature of Aboriginal population distribution is the "outstation" movement, which began in the early 1970s and has expanded considerably since then. Small groups of Aborigines have been moving away from larger settlements to bush locations, in an effort to reaffirm links with their land and their culture. At outstations some people hunt and gather, live in bush shelters, and participate fully in rituals. However, most outstations include considerable European technology, especially in terms of transportation and communication. More than 20,000 Aborigines are involved in the movement, using some 300 outstations, mostly in the Centre and Top End of the Northern Territory.

In contemporary Australia, two issues—land rights and substance abuse—are of overwhelming importance in Aboriginal society:

1. **Land rights.** Private land ownership was unknown in pre-European days. Today most of inland Australia still is Crown land (government-owned), and is only leased to pastoralists or other users. In recent years, however, various Aboriginal groups have demanded that they be granted both ownership of tracts of "traditional" land and the underlying mineral rights (minerals are

[5] Prior to 1966 Aborigines were not counted in the Australian quinquennial census of population. Enumeration improved with each successive census, but the data are still considerably less reliable than those for the rest of the population. Since 1986, aboriginality has been self-declared in the census. There is a single question: "Do you class yourself as Aboriginal, and would you be accepted by the Aboriginal community?"

[6] In 1984, a group of nine Pintubi people came out of the Great Sandy Desert of Western Australia and saw whites for the first time. The are considered to be the very last of the pure desert dwellers, and perhaps the last people in the world to have been completely isolated from civilization.

the property of the government throughout Australia). In 1977, the federal government began granting Aborigines freehold title to existing reserve land and some other vacant Crown land in the Northern Territory (the federal government has jurisdiction over land only in the territories, not in the states). Titles are held by Aboriginal Land Trusts and administered by Aboriginal Land Councils.

At the time of writing, about 45% of the total area of the Northern Territory is in Aboriginal ownership. South Australia passed similar legislation, and about 10% of that state's area (albeit the most arid portion) is now owned by Aborigines. No other state has vested a significant amount of land to Aboriginal ownership. In the Northern Territory and South Australia mineral rights were not transferred to the Aborigines, but their consent must be obtained prior to exploration or mining, which gives then *de facto* control and significant royalty payments.

Of potentially much greater significance is a 1992 ruling by the Australian High Court (after 10 years of litigation) that affirmed that the Meriam people (a Torres Strait Islander clan) had full legal native title to a small island (Mer or Murray Island) where they and their ancestors had lived for centuries. This judgment, called the *Mabo Case*, after the leader of the plaintiffs, thus invalidated the long-held concept of *terra nullius* (land belonging to no one), a legal tenet that maintained that no one had laid claim to the continent before the British did. The Aborigines, due to their lack of social organization (in British terms) had been deemed not to have any proprietary rights of title.

The Mabo decision denied *terra nullius*, and indicated that Aborigines had rights to common law native title if they could prove that they had continuously maintained their traditional association with the land claimed, and if those rights had not been previously extinguished by government action. The immediate result of this court decision was massive confusion. Some Aboriginal groups laid extravagant claims (e.g., to downtown Brisbane or to most of southwestern Queensland), mining companies predicted that their industry would collapse, politicians ranted about the possible take-over of suburban homes, etc.

Undoubtedly, there will be several years of legal wrangling before the implications of the Mabo decision are ascertained. The limitations of the judgment are such that it will apply mostly to vacant Crown lands, lands reserved for public purposes (such as national parks), and some lands now held in leasehold. Meanwhile, widespread uncertainty and bewilderment will affect many aspects of Australian economic, social, and political life.

2. Substance abuse. Aboriginal leaders are in virtually unanimous agreement that chronic substance abuse, primarily alcohol, is their biggest social problem. Many Aboriginal settlements, including more than half of those in the Northern Territory, have banned the sale of alcoholic beverages, but there is also a reaction against this as a paternalistic attitude.

CHAPTER FOUR

Fruits of the Land: The Primary Industries

We have seen that Australia is a vast land with a sparse population, situated in a relatively remote part of the world. It has an advanced capitalist economy and a high standard of living, both of which are based upon a mixed, but generally well-endowed, array of resources, a well-established transportation and communication infrastructure, an urban-oriented society, a capitalistic economic system, a skilled and energetic populace, and liberal infusions of Western technology and capital.

The Aboriginal Economy

For many hundreds of generations, the economic development of Australia was totally in the hands of the Aborigines, who were its only human inhabitants. Theirs was a simplistic economy, based upon hunting, fishing, and gathering. It encompassed no crop-growing, no livestock, and only rudimentary trade. It used nature's provender to the utmost, and appeared to be reasonably well-balanced with the environment in ecological terms.

The Aborigines were nomadic, although their wanderings stayed within territorial limits. Only in certain water-side locations were there any long-term settlements or any significant population density. Most of the hunting, especially for "big meat" (kangaroos, wallabies, larger birds, etc.), was done by the males, primarily with spears, but also using snares, boomerangs, and other implements in some situations. Women and children generally provided the bulk of the day-to-day food with their gathering activities, seeking lizards, eggs, invertebrates, seeds, nuts, roots, and other vegetable matter. Fishing was a variable but very important activity for those groups located along a seacoast or a stream; they used spears and poisons and were clever in constructing fish traps.

The Australian environment was not particularly fruitful for a hunting/gathering economy, except in a relatively few locations. Even so, it provided adequately for a scattered Paleolithic population for many thousands of years.

The Resource Base

At first glance the resources of Australia are somewhat less than impressive. The fundamental failing is a paucity of water, in both precipitation and streamflow. No other inhabited continent is so dry, and the influence of this aridity is clearly felt in the land-based primary industries. The arid milieu is conspicuously reflected in the landscape by the pattern of the natural vegetation. Forests are found in only a few areas, and there is a particular scarcity of softwoods. Lesser plant associations are dominant, and their presence emphasizes the negative aspect of the resource base. The soil resource comprises another problem; the prevalence of impoverished soils is accented by a lack of nutritive trace elements and an excess of salinity. Even the oceanic fishery has been unimpressive; despite the fact that fish comprise a minor part of the Australian diet, the country has been a net importer of fish products throughout its history.

There is, however, another side to the coin, and the brighter aspects of the resource base are becoming increasingly evident. First and foremost are the abundance and diversity of minerals of economic value. Probably only two other countries in the world, the United States and Russia, possess a quantity and variety of ores greater than that of Australia. Indeed, Australia is among the world's leaders in output of every major industrial mineral except petroleum and natural gas.

Three other felicitous aspects of the Australian resource picture are forage, soils, and fisheries. A continent with limited forests is likely to have extensive areas of grass and shrub that can support a grazing industry. Such is the case in Australia, where the superb natural forage value of such species as Mitchell grass, kangaroo grass, and some types of saltbush and bluebush has been augmented by the introduction of a number of exotic pasture grasses. Although Australia's soils are not notably fertile, many of them are quite responsive to the addition of chemical fertilizers, especially phosphate and various trace elements, which have raised productivity levels significantly in many areas. Only in the last few years has there developed an appreciation of the potential fishery resources in the seas surrounding Australia. While not yet exploited with much élan by the domestic fishing industry, the results of ingenious exploitation by foreign fishermen have not gone unnoticed, and marine resources are playing an increasingly important role in both the economy and diet of Australians.

In a spacious land with an uneven endowment of natural resources and a small population, the "development" of these resources is a matter of

Focus Box: The Snowy Mountains Scheme

Australia is the locale of one of the world's most ambitious water storage and diversion enterprises, the *Snowy Mountains Scheme*. It is a dual purpose (hydroelectricity generation and irrigation) project that was begun in 1949. Construction spanned a quarter of a century and was completed in 1974. The general plan was a simple one, but its implementation was difficult and expensive.

The eastern and southern slopes of the Snowy Mountains in southern New South Wales, the highest part of the Eastern Highlands, receive considerable year-round precipitation. The rivers that drain this area southeasterly into the Pacific Ocean carry a great deal of water that is unused by people and thus flows into the sea as a "wasted" resource. The heart of the Snowy Scheme is the impounding of two of these rivers—the Snowy and the Eucumbene—and diverting much of their flow through trans-mountain tunnel systems to the west slope of the range. This water is used to generate hydroelectricity and is then added to the Murray and Murrumbidgee Rivers, where most of it is used for downstream irrigation projects.

Major construction included 17 large dams and many small ones, nine power stations, about 100 miles (160 km) of tunnels, 80 miles (128 km) of aqueducts, and many miles of high-voltage electricity transmission lines. The cost of the scheme has been enormous (roughly one billion dollars), but its results are impressive. The average annual electricity output is more than 5 billion kWh, and the average annual addition of irrigation water amounts to some 600 billion gallons (2,300 Gl).

great concern, and judicious application of the relatively limited amounts of investment capital is called for. Nevertheless, development schemes, financed in both the public and private sectors of the economy, have been advanced on many fronts. The impoundment and diversion of surface waters has attracted most attention in the past, and will continue to loom large in the plans of resource developers at all levels. The most conspicuous of these efforts has been the Snowy Mountains Scheme, a billion-dollar project to divert the waters of two east-flowing rivers (Snowy and Eucumbene) into the west-flowing River Murray system for the purpose of hydroelectricity generation and increased irrigation usage. This 25-year project was completed on schedule, and, although its cost/benefit ratio is questionable, it has become a national status symbol of utmost significance.

Soil improvement by mineral application has provided a less spectacular but no less impressive expression of resource development. For example, the fertile South Australian farming area now called Coonalpyn Downs was known prior to the 1950s as the 90-Mile Desert, but minute additions of zinc and copper to the soil transformed an unproductive region into a major producer of grains, wool, and fat lambs. Forage improvement has also yielded fruitful results; notable projects currently under way include the large-scale eradication of brigalow scrub (*Acacia harpophylla*) in areas of central and southern Queensland to promote the growth of more nutritious forage, and widespread scrub clearance and pasture improvement in the remote Esperance region of Western Australia.

An important botanical immigrant has been the exotic softwood; Australia has no indigenous pines (*Pinus* spp.), but a number of Northern Hemisphere pines have been established in plantations in various localities, giving rise to both lumber and pulp industries of significance. These are but a few examples of resource development projects, representing a field of endeavor that continues to occupy much attention among Australians. The country is fortunate in having a well-established federal research organization, the CSIRO (Commonwealth Scientific and Industrial Research Organization), which has been a world leader in applying the results of scientific research to the solving of practical resource development problems.

Extractive Industries

Mining: Dynamism for Today and Tomorrow. The extraction and processing of minerals have been major factors in the economic development of Australia, almost from the time of earliest European settlement. A wide variety of minerals is found in commercial quantities, and in recent years Australia has become one of the world's leading mineral exporters. The total complement of mineral resources gives Australia an attractive base for industrialization. Furthermore, the establishment of mining centers, often in remote localities, has had a salutary effect on the spread of settlement, and some of the mineral discoveries have sharply stimulated immigration into the country.

Mining influences the nation's economic health out of all proportion to its size. It employs only about 1.5% of the work force, and accounts for about 6% of (GDP) Gross Domestic Product. However, about two-thirds of all mineral output is exported, and mining generates about one-third of total export income.

Coal has been the keystone of the mineral economy. It was the first mineral to be mined (in 1796) and the first to be exported (1901), and coal-mining still employs one-third of all the miners in the country. Australia is the world leader in coal exports, shipping about half the total amount of coal that is traded internationally. Every state except Tasmania has significant coal production, but more than half the output is from three areas in New South Wales: the northern coalfields in the lower Hunter Valley, the western coalfields inland from Sydney around Lithgow, and the southern coalfields in the vicinity of Wollongong (see Figure 4-1). These fields, approximately equidistant from Sydney, are the most significant in Australia from the standpoint of quantity, quality, and accessibility. Queensland ranks second among the states in both production and reserves; in recent years there has been rapidly increasing production, especially for export to Japan. In addition to the major uses of coal for heavy industry and railways, the generation of thermoelectricity provides a notable market for Australian coal. For example, essentially all of the commercial electricity in South Australia is generated

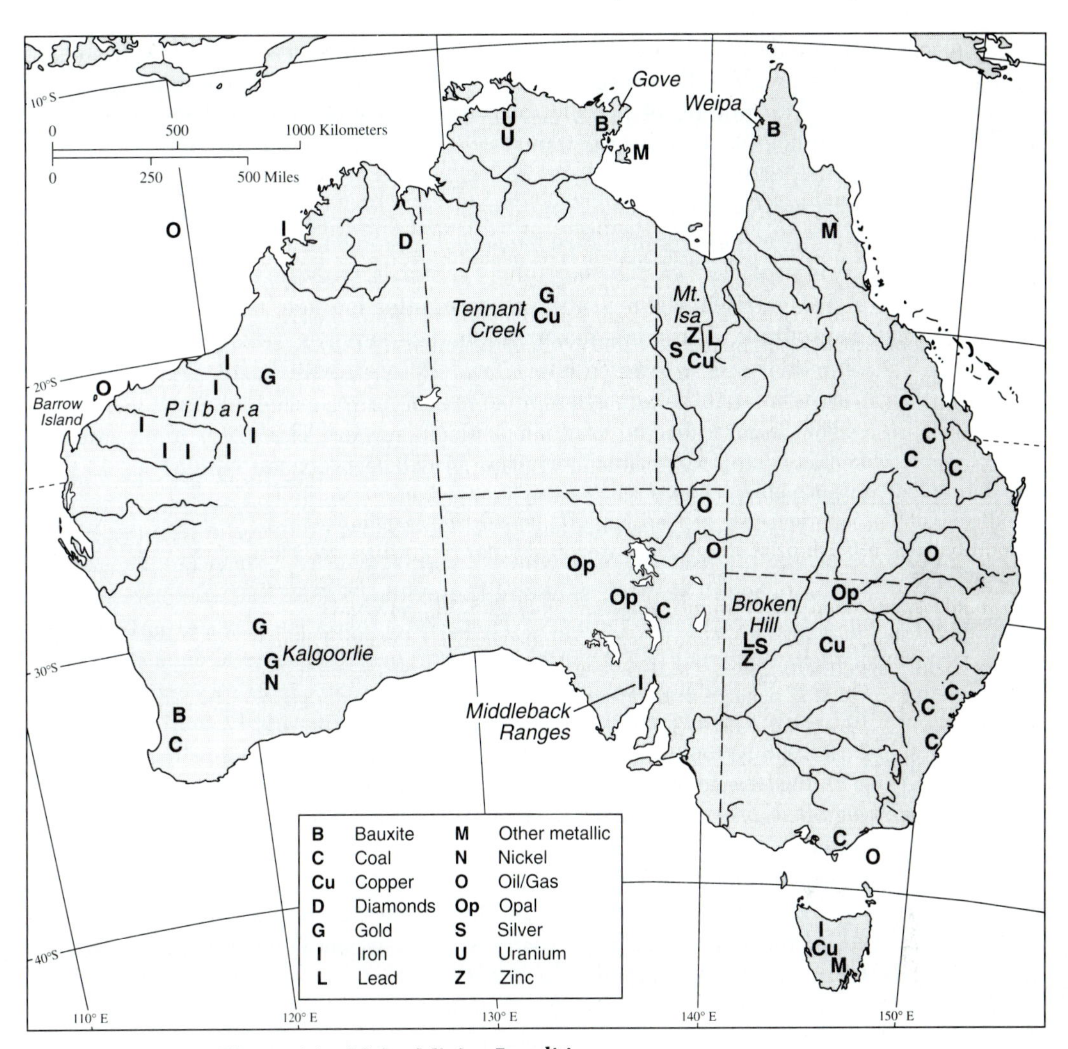

Figure 4-1 Major Mining Localities

from subbituminous coal, and most of Victoria's electricity is produced from enormous deposits of lignite (brown coal) in the Latrobe Valley, east of Melbourne.

If coal has been the keystone of the Australian mineral industry, *gold* has been the catalyst. As was mentioned in Chapter 3, early gold rushes in Victoria and New South Wales had a pervasive influence on the economy and settlement patterns of those two colonies, and the later strikes in Western Australia were equally important. Gold continues as a major mineral product, In the early 1990s, gold was the second largest export earner after coal, with Australia producing about ten per cent of the world's gold output. Principal production is from the long-standing Kalgoorlie area of Western Australia, although new mines have been opened and some old ones re-opened at various locations since gold prices skyrocketed in the 1970s.[1] Moreover, considerable gold is recovered as a by-product of base metal mining.

The mining of *nonferrous metals* is a more recent development in Australia, but much of the nation's reputation in mining has come from this branch of the industry. *Lead* and *zinc* ores normally occur together, and must be separated in concentrating plants; *silver* is usually produced as a by-product of the refining of lead and zinc. Australia is among the five leading world producers of all three of these minerals, with the bountiful ore bodies of Broken Hill (New South Wales) accounting for more than three-fourths of total output. The other major production center is Mt. Isa (Queensland). *Copper* is also a major product from Mt. Isa, with Roxby Downs (South Australia) as the other major producer. *Bauxite* is another Australian specialty; it is by far the world production leader. The major mines are at Weipa on the western side of Cape York Peninsula and Nhulunbuy in the northeastern corner of Arnhem Land.

Historically Australia has been a moderate producer of *iron ore*, with steady output from the Middleback Ranges in South Australia and the islands of Yampi Sound in Western Australia providing a more than adequate supply for the domestic steel industry.[2] However, a relaxation of ore export restrictions in the mid-1960s exposed a vast cornucopia of mineral deposits, especially in the Pilbara District of Western Australia. Almost overnight some of the world's largest reserves of iron ore were "discovered," and a frenzy of activity (developing mines, laying railway track, dredging new harbors, and building new towns) has transformed the somnolent Northwest into a boom region of great economic excitement (see Photo 4-1). Japanese capital was prominent in the development of the Pilbara deposits, and all of the ore is exported, mostly to Japan.

Even more important in the long term may be the development of a *petroleum* and *natural gas* industry, which was completely lacking before the 1960s. Australia's first oil field was brought in near Moonie in southeastern Queensland in 1961; since then commercial production of both oil and gas has begun at several other localities, particularly in Bass Strait off the Gippsland coast of Victoria. Australia is now more than 60% self-sufficient in petroleum resources (lacking mostly heavy crudes). Moreover, it produces natural gas sufficient to supply domestic demand, and has reserves that far exceed domestic needs. The most dynamic recent development has been in the so-called North-West Shelf area, which is well offshore (180 miles or 290 km on average) from the coastline of northwestern Western Australia. Both oil and gas are now being produced commercially there, and a major export is *liquified natural gas* (LNG), mostly to Japan.

As overwhelming as the iron and petroleum discoveries have been,

[1] In the early 1990s the two leading gold mines in Australia were the "Super Pit" at Kalgoorlie and the Telfer mine on the edge of the Great Sandy Desert in the northwestern part of Western Australia.

[2] After a half century of production, the iron mines of Yampi Sound finally were closed down in 1993.

Photo 4-1 *About a dozen huge open-pit iron ore mines have been developed in the Pilbara district of Western Australia since 1960. This is the Mount Whaleback mine near Newman. (TLM photo.)*

they are not the whole story. A varied storehouse of minerals is being exploited, including some of the world's largest deposits of *uranium, nickel, manganese, mineral sands*, and *diamonds*.

The mining breakthrough of the 1960s must rank with the introduction of Merino sheep and the development of refrigerated shipping as one of the most significant events in the entire economic history of Australia. It is likely that more discoveries, perhaps *many* more, will be made before the activity subsides into a state of steady productivity, although the uncertainties surrounding the Mabo decision are likely to delay some potential development.

Logging: Little Today and Less Tomorrow. It has already been noted that Australia is poor in timber resources because of the great extent of treeless areas and because of the limited varieties of timber available in those areas that are forested. Only about 2% of the country is classified as "accessible" forest land; the comparable figure in the United States is 22%. Almost all the indigenous timber cut in Australia is hardwood, the bulk of it eucalyptus. Exotic softwood plantations, largely of Monterey pine (*Pinus radiata*) from

California, have been developed in every state; these somewhat assuage the acute shortage of native softwoods.

Each of the states has a timber industry of some importance, but the largest sawmill output is in New South Wales and Victoria. In the last few years a significant woodchipping industry for pulp and paper manufacturing has developed in Tasmania, New South Wales, and Victoria. Clearcutting practices are employed, and the devastated landscape that results has become a rallying point for proponents of preservation/conservation.

As the Australian population continues to increase, there is a concomitant growth in demand for wood products, but with little hope for expansion of domestic supply. It seems inevitable that the proportion of imported wood and wood products will continue to rise.

Fishing: An Unrealized Potential. Although Australia has traditionally been oriented toward the sea for its commercial links with the world, relatively little attention has been paid to maritime fishery resources until recently. Fish have been unimportant in the diet of most Australians, and there has been little attempt at export of fishery products. For the most part, the function of the fishing industry has been to supply fresh fish to the local metropolitan markets, and the only important fishery exports have been those (such as mother-of-pearl, pearl shell, and trochus shell) derived from the mollusc fisheries in the tropical waters off northern Australia.

This pattern began to change during the 1950s, when an important tuna fishery in southern waters began to be exploited, especially off the coast of South Australia. Further diversity was introduced in the 1960s with the rapid development of a crayfish (lobster) industry off Western Australia; the export of frozen cray tails to the United States burgeoned into an important source of foreign exchange. Under the stimulus of investment and expertise from Japan, cultured-pearl fisheries have been initiated in more than a dozen sheltered bays of the north coast; commercial oyster farming is carried on in several localities along the coast of New South Wales; and an expanding prawn fishery is being developed in a number of places.

The government attitude toward commercial fishing has become more aggressive in recent years, catalyzed by the continued successes of Japanese and Russian fishermen off the Australian coast. The proclaimed limit of Australian jurisdiction has been extended from three to 200 miles (4.8 to 320 km) offshore, and more financial and legal help has been granted to the fishing industry. Even so, it should be kept in mind that fishing holds only a minor place in the Australian economy; for example, the total number of people employed in commercial fishing is less than the number of lead-zinc miners in the country.

Agricultural Industries

Throughout its history Australia has been heavily committed to the agricultural sector of the economy, both crops and livestock. Even in the atomic age, with most Australians living in cities and the great bulk of the labor force employed in non-rural occupations, the "primary industries" continue to play an important role. The nation is almost self-sufficient in agricultural produce; the only farm products of significance that must be imported are tea, tobacco, chocolate, and coffee. Moreover, agricultural exports have long been significant to the national economy, although their relative importance continues to decline.

Crop Farming: A Question of Markets. Probably the most remarkable thing about farming in Australia is the limited portion of the continent that is involved. Only a small fraction of the land area has the proper combination of climate, terrain, and soils to permit crop growing, and that small area is very restricted in its location. The major agricultural zone, within which farming appears in discontinuous segments, extends around the southeastern margin of the country in the form of a crescent that is from 100 to 250 miles (160 to 400 km) wide, with its northeastern extremity in the Darling Downs of southeastern Queensland and its western end on the Eyre Peninsula of South Australia (see Figure 4-2). There is a comparably placed, though smaller, region occurring as a northwest-southeast trending band on the subhumid margin of Western Australia. Smaller farming areas, some of which are quite significant, are found in:

1. Some of the coastal lowlands and river valleys of eastern Queensland, from Brisbane to Cooktown
2. The north coast of Tasmania
3. The southeastern portion of Tasmania
4. An area extending inland from Esperance on the southern coast of Western Australia
5. A limited area around Perth in Western Australia

Despite the relatively small amount of land devoted to crops, however, it is more than sufficient to supply the domestic demand for foodstuffs and fibers. Greater acreages could be made available for most types of farming if the market were larger. Although there are occasional crop failures, the prime determinant of total production is normally the ability of Australian produce to penetrate overseas markets.

Wheat. Wheat is easily the leading crop; it occupies nearly half of all the cropped acreage and accounts for nearly one-third of the total value of all crops (see Table 4-1). For the most part wheat is grown as a mixed farming enterprise, with sheep raising as the other important element. There is usually a distinct correlation between the acreage and relative market prices for wheat and wool; as the price for one goes up, the acreage devoted to the other is likely to be curtailed. Thus, high wheat prices encourage the farmer to sow more acres to that crop and decrease the acreage allotted to sheep

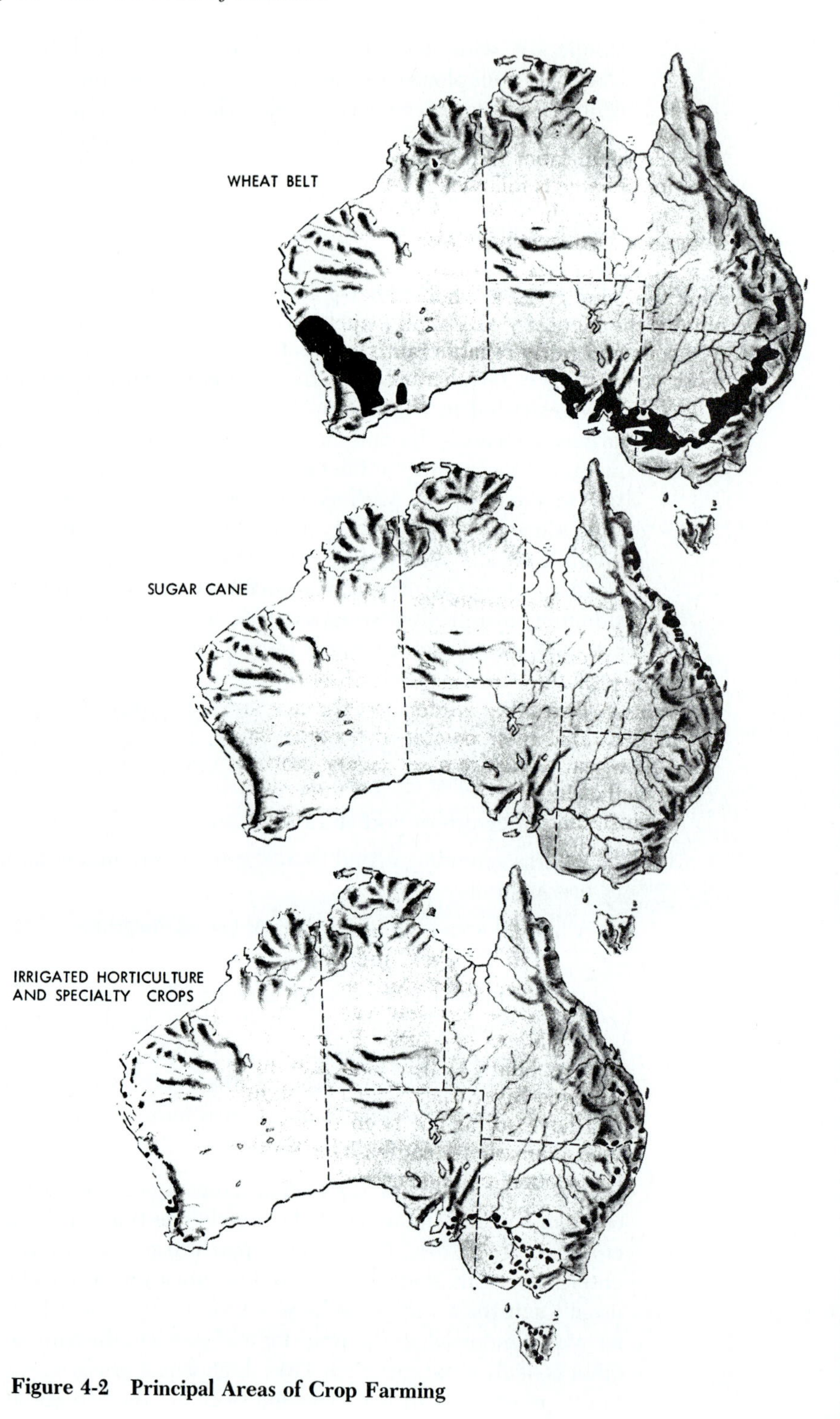

Figure 4-2 Principal Areas of Crop Farming

Table 4-1. *Australia's Leading Crops (Proportion of Total Gross Value of Output, 1989–1990)*

1. Wheat	28%
2. Sugar cane	9%
3. Cotton	7%
4. Barley	7%
5. Hay	5%
6. Grapes	4%
7. Potatoes	3%
8. Apples	2%
9. Bananas	2%
10. Oats	2%
11. Tomatoes	2%
12. Oranges	2%

pasturage, and *vice versa*. For example, in the early 1990s, wool prices were severely depressed, resulting in a marked decline in sheep numbers and concomitant increase in planting of wheat and other grains.

The basic environmental association is with precipitation; most wheat production is on lands that are semiarid to subhumid. In Western Australia, South Australia, and most of Victoria wheat is grown in areas that receive between 10 and 20 inches (250 and 500 mm) of annual rainfall, with the 20-inch isohyet generally marking the seaward margin of the wheat belt. In New South Wales, wheat is grown between the 15-inch (375 mm) and 30-inch (625 mm) isohyets, and it is grown in areas of slightly higher rainfall totals in Queensland.[3]

Australian wheat farming is an extensive operation, involving large farms with a relatively large capital investment, much use of machinery, and a relatively small labor requirement. Most of the wheat is grown on flat or undulating country inland from the coastal highlands. This permits the easy use of large-scale machinery at all stages in production from soil preparation to harvesting. In most wheat areas, a four- to six-year rotation scheme is followed, alternating with pasture, other grains, and fallow. Superphosphate is used liberally as a fertilizer, and in recent years nitrogenous fertilizers have also been heavily used. Even so, average yields are relatively modest. The Wimmera District of Victoria usually obtains the highest yields of all major producing areas, on account of its fertile soils and fairly reliable rainfall (see Photo 4-2).

[3]Students of American geography will note that wheat farming in this country occurs in areas of greater annual precipitation; the apparent discrepancy can be resolved by a consideration of winter precipitation totals. Wheat is a winter crop (entirely in Australia; mostly in the United States), which means that it is planted in the fall and harvested in early summer. The Australian wheat belt receives most of its annual rainfall during the winter (which is the growing season for wheat), whereas most American wheat is grown in areas that have a summer precipitation maximum. Thus, higher annual totals in the United States are offset by a greater proportion of winter rain in Australia, which generally equalizes the growing season rainfall in the two countries.

Photo 4-2 *A typical November wheat scene in the productive Wimmera district of Victoria. A rich stand of unripened grain in the foreground contrasts with the storage silos on the horizon. (TLM photo.)*

In most years, the greatest acreage and production of wheat is in Western Australia, with New South Wales also recording large totals. South Australia is a strong third, and Victoria is also important. Wheat output in Queensland and Tasmania is much more limited.

A federal agency, the Australian Wheat Board, administers all transportation and marketing arrangements for wheat. The agency provides price supports loosely based on world market prices, but cushioning any abrupt price declines. As a fixed price is not guaranteed, it encourages farmers to weigh opportunities for alternate commodities.

Wheat continues to amount to more than half of all grain production, being four times as important as the second-placed grain (barley).

Other Grains. Two other winter cereals, oats and barley, are grown extensively in Australia. *Oats* comprises a secondary crop grown throughout the wheat belt, as well as in some of the wetter areas in southeastern Australia where wheat is insignificant. Oats serves as a triple-purpose crop; it is grazed by livestock, cut for hay, and harvested for grain. Eventually most of it is fed to livestock, with about one-fourth of the total crop being exported.

Barley replaces or complements wheat in certain parts of the wheat belt, particularly in South Australia. Approximately one-third of the total crop is exported, one-third is fed to livestock, and one-third is malted or distilled.

There are three summer grain crops of significance. *Rice* output has increased dramatically in the last few years, with a large share of the crop exported to Papua New Guinea and other Pacific islands. Production is highly mechanized and yields are prodigious. In some years, Australia has the highest per-acre production average in the world. Most rice-growing take place in the MIA, but it is cultivated in several other irrigated valleys in New South Wales and Queensland. Aspirations for significant rice production in tropical Australia have so far not been realized. Output of *grain sorghums* has also been increasing steadily. The leading production area is the Darling Downs section of Queensland. *Maize* (corn) is grown for grain in Queensland (especially the Darling Downs and the Atherton Tableland), and as a fodder crop associated with dairying in other parts of the eastern states.

Sugar Cane. Australia's leading specialty crop is sugar cane, which has been grown with considerable success for more than a century. It is produced on some 8,000 farms, mostly small, on the discontinuous coastal lowlands of Queensland and northernmost New South Wales. Much more cane could be grown with comparatively little effort, but the international market is very unreliable. Production quotas, as well as prices, are tightly controlled by the Queensland Sugar Board, which handles all marketing arrangements. Since the 1960s, when Cuba lost much of its international market for sugar, Australian production has increased significantly. Australia is now one of the five leading cane producers in the world, and is second only to Cuba as a sugar exporter, with major sales to the United States, Japan, and the United Kingdom. Nevertheless, the industry is now in great difficulty, due to incautious expansion in the 1970s. Many farms have failed, and many other farmers are deeply in debt.

Horticulture. The intensive growing of fruits and vegetables occupies only about 1% of Australian cropland, but comprises nearly one-fourth of total crop value. The output of almost every horticultural crop could be expanded considerably if there were markets available. Each state has an *apple*-growing area in a temperate climate zone near the capital city; most exports emanate from Tasmania, where apples are a major crop. Other deciduous fruits, especially *pears*, *peaches*, and *apricots*, are notable crops in various irrigated valleys, particularly those of the Murray, Murrumbidgee, and Goulburn rivers. Much of the output of these fruits is canned or dried for export. *Oranges* are grown on sandy soils, usually under irrigation; major areas of production are the Gosford district just north of Sydney, the Murrumbidgee irrigated areas, the Murray irrigated areas, and the hills around Perth and Adelaide.

Three tropical fruits are produced in large volume along the east coast and in much smaller quantity in Western Australia. The principal area of *pineapple* production is in the low hills abutting the coast of southernmost

Queensland, near Brisbane. *Bananas* could be grown widely in the high-rainfall areas of coastal Queensland, but most production is concentrated along the extreme north coast of New South Wales. In the last few years, there has been a notable expansion of banana growing in the north Queensland (Cairns-Townsville) coastal belt. Most of the Perth market for bananas is supplied by irrigated farms near the mouth of the Gascoyne River at Carnarvon, but there has been expanded recent production in the Ord River irrigation project near Kununurra. Specialized growing of *mangos* has suddenly become a major enterprise in the Cairns-Townsville coastal country of Queensland, and mangos are also increasing in importance in the Carnarvon and Kununurra districts.

Most *vegetables* are grown in market gardening situations near the larger urban centers, to take advantage of the local market for fresh produce. In many cases the soils in such localities are only moderately fertile, but hard work and the use of fertilizers make them productive. *Potatoes*, on the other hand, need a cool, moist environment, so they are mostly grown in Tasmania and the cooler parts of Victoria, although there have been significant recent acreage increases in New South Wales.

Viticulture. There is a notable amount of grape growing in Australia, much of it dating from the early days of settlement. Only about 10% of the crop consists of table grapes; the balance is evenly divided between wine and drying varieties (see Photo 4-3). The South Australian vineyard areas, of which the most important is the Barossa Valley north of Adelaide, emphasize grapes for wine and brandy. The irrigated areas of the Murray and Murrumbidgee valleys specialize in varieties that can be dried into raisins, sultanas, and currants.

Nonedible Crops. Two specialty crops are particularly notable in Australia. Production of *cotton* has demonstrated a remarkable increase in recent years. Many areas are environmentally well-suited to cotton growing; all important cotton areas use irrigation. The Namoi district of New South Wales, where production was initiated by immigrants from California and Arizona, is easily the most important. By the early 1990s, Australia had become the fourth largest exporter of cotton in the world. Although *tobacco* growing is diminishing, due to withdrawn government price supports and a downturn in smoking, it is still widely grown in Australia. Principal production areas are in Queensland (especially the Atherton Tableland and the lower Burdekin Valley) and Victoria (primarily the Ovens Valley).

Sheep Raising: The Dominant Rural Industry. Despite Australia's diversified crop production, until recently, animal husbandry has been more important than farming. At the time of the writing, the value of livestock products is about equal to that of all crops.

Ever since the early days of settlement the Australian economy has been said to "ride the sheep's back." Farming was difficult in the sterile sandstone hills around Sydney, and the early colonists soon turned to sheep

Photo 4-3 *There are many vineyard areas in Australia. This scene is just northeast of Perth, Western Australia, with the Darling Scarp in the background. (TLM photo.)*

raising as a source of income. In fact, most exploration outward from Sydney and other coastal settlement nodes was at least partially motivated by the desire to find more and better sheep pasturage. The emphasis has always been on wool production, so most of the early sheep imports were varieties of Merino. The present-day Australian Merino has been bred from Spanish, South African, English, French, and German stock, and it differs considerably from its ancestors. It is relatively large-framed and hardy and has proved to be adaptable to an arid climate as well as to high and low extremes of temperature. About three-fourths of all sheep in Australia are Merinos or Merino crossbreeds.

Australia contains nine times as many sheep as it does people, the total of about 150,000 amounting to nearly one-fifth of all the sheep in the world. Australia's wool output is 30% of the world total, which is more than twice as much as the second-ranking producer. Although sheep raising has been tried wherever conditions held out any hope for success (and in many places where they didn't), most Australian sheep are found in the southeastern and southwestern portions of the country, generally in a broad crescent along and inland from the "wheat belt." The greatest concentrations are in New

South Wales (more than one-third of the total), where sheep raising dominates the rural scene almost everywhere except within 50 miles (80 km) of the coast. Western Australia and Victoria each have about one-fifth of the Australian sheep population; South Australia and Queensland have about 10% each; Tasmania has a smaller proportion; and there are no sheep in the Northern Territory.

Extensive Wool Production. There are three basic types of sheep raising operations, of which the most widespread is the extensive pasturing of Merinos for wool production. As is seen in Figure 4-3, this type of operation occupies broad areas in the interior of Queensland, New South Wales, South Australia, and Western Australia. Large landholdings are typical, ranging in size from only 1,000 acres (400 ha) in better-watered areas to more than 800,000 acres (320,000 ha) in drier regions. Carrying capacity is low; in some cases as many as 25 or 30 acres (10 or 12 ha) are required to support one sheep, and provision of watering points is essential. About half of these properties carry between 1,000 and 5,000 sheep each, but 40% of them involve more than 5,000 animals each. Altogether, about one-fourth of the national sheep flock is pastured in this zone.

These extensive properties, called "stations," usually breed their own stock, maintaining a nucleus of stud rams along with large numbers of ewes. The sheep are allowed to shift for themselves within large fenced pastures ("paddocks") most of the time. They are rounded up ("mustered") once a year, usually in fall or winter, for shearing, which is accomplished by an intinerant shearing team in a shearing shed that is often the most imposing building on the property (see Photo 4-4). On the larger and more remote stations the annual shearing muster is the only time during the year when the entire flock ("mob") is brought under systematic scrutiny. On smaller and more heavily stocked properties, however, there may be other musters, at which time the animals are given such special care as spraying or dipping to combat ticks and inhibit flies.

Although the continent is unusually free from diseases that affect sheep, the stockman must contend with a number of other environmental problems. *Drought* is the most notable, of course; extended periods without rain cause both a decline in lambing percentages and an actual die-off in adult sheep, because on these extensive stations it is impractical to give the animals special care and they must prosper or decline on the basis of their ability to cope with nature. *Parasites*, both internal and external, sometimes cause significant problems. *Vermin*, comprising animals that compete with sheep (such as kangaroos and rabbits) or prey on sheep and lambs (such as dingoes and foxes), play a leading role in Australian demonography. Extraordinary efforts have been expended to eliminate them from sheep areas, including the construction of tens of thousands of miles of rabbit and dingo fences.[4]

[4]One dingo barrier fence extends unbroken through three states for more than 5,300 miles (8,500 km).

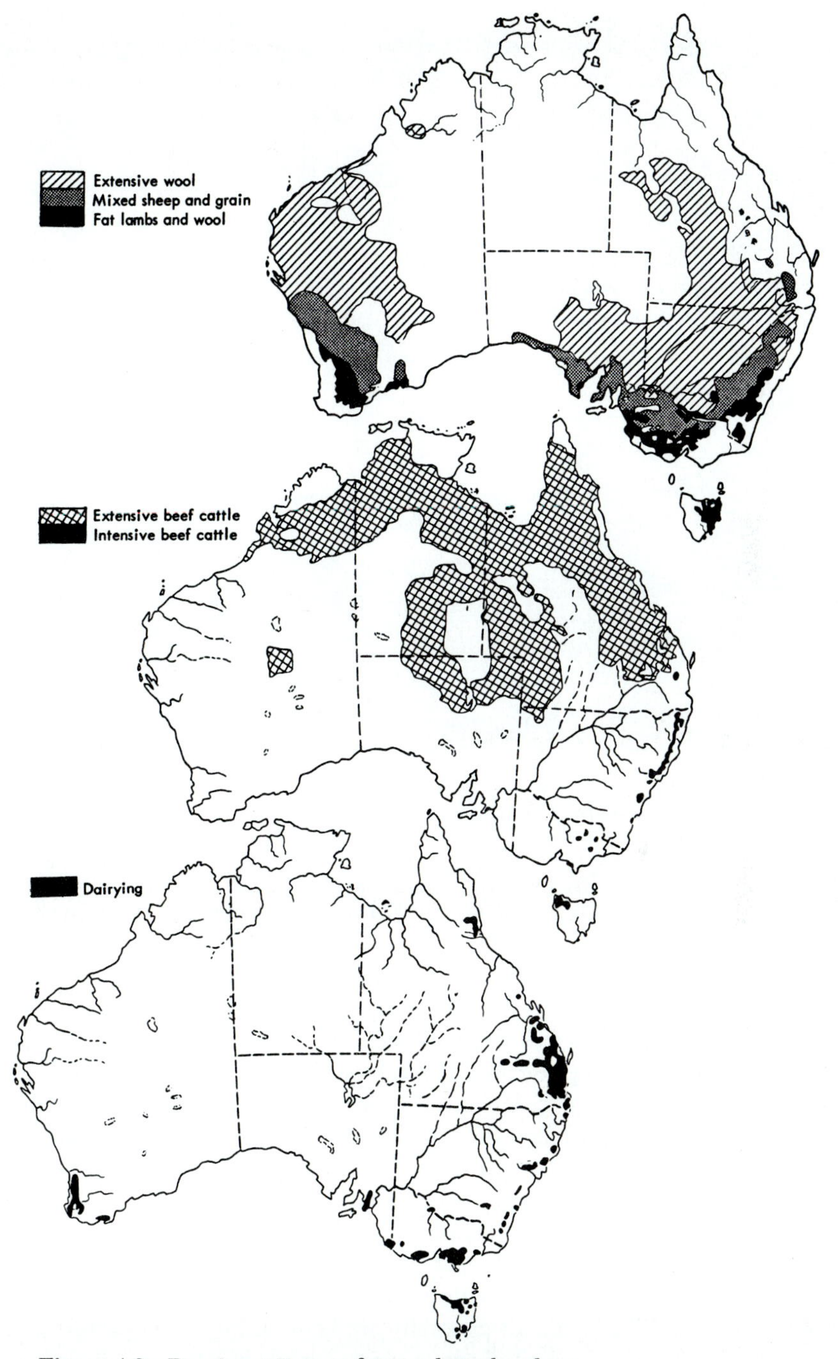

Figure 4-3 Dominant Types of Animal Husbandry

Photo 4-4 *Heavily fleeced Merino sheep waiting to be shorn, near Armidale in northeastern New South Wales. (TLM photo.)*

Mixed Sheep and Grain Farming. Throughout the wheat belt there are many properties that produce both crops and livestock products for sale. Wool and wheat are the principal products from these farms, but barley and oats are also raised in considerable volume, and the farmer sometimes emphasizes the output of meat (lamb or mutton) rather than wool.

Sheep are more closely supervised and better cared for in areas of mixed farming, as both the size of the property and the number of animals are usually much smaller. In normal times there is a considerable movement of sheep from the extensive wool areas into the mixed farming zone. This involves the surplus stock from the wool properties, and mostly consists of old ewes (female sheep) and wethers (castrated male sheep). The farmer can take these animals and obtain a wool clip, then fatten them for sale, or sometimes breed them to produce fat lambs to sell.

The animals generally are kept on improved pastures, supplemented by some grazing of the growing cereal. Flocks typically are a few hundred in size, but often a few thousand. Approximately one-half of all Australian sheep are involved in these mixed farming operations.

Fat Lamb Production. The third principal type of sheep husbandry involves the production and fattening of lambs to sell for meat. Coarse wool (non-Merino or shorter wool) is usually a by-product of these operations. This is a fairly intensive form of pastoralism, involving improved, high quality pasturage. As demonstrated in Figure 4-3, the fat lamb areas are in the moister southeastern and southwestern corners of the continent, as well as in Tasmania. Better meat producing varieties of sheep, such as the Polwarth (a breed developed in Australia), Leicester, Suffolk, and various crossbreeds, are used. Flock size is relatively small, generally a few hundred animals, but one-fourth of the national sheep total is found in this zone.

Recent Trends. In the late 1980s and early 1990s, Australian sheepmen faced one of the worst economic crises in history. Due to oversupply and weak demand, Australian wool prices fell below the price of production. Wool export income declined by about half between 1988 and 1992, with the average producer losing more than $30,000 in the latter year. Part of this decline was due to the economic problems associated with the fragmentation of the Soviet Union and the demise of Communism in Eastern Europe, which effectively removed a large share of the market for Australian wool. Most Australian wool is marketed by auction, and in 1993 there was an unsold stockpile of nearly four million bales. Whether this is a long- or short-term economic dislocation is still to be seen.

Beef Cattle: Increasing in the Outback The early husbandry of beef cattle was not as widespread as that of sheep. Generally beef cattle raising has been developed in areas not suitable for the more profitable sheep raising and dairying, although beef cattle are an adjunct to both in moister country. In the northern portion of the continent there was a more distinctive and clear-cut development of cattle raising. Western and much of northern Queensland had become cattle country by the 1860s, and there was substantial cattle interest in the following decade in the Northern Territory. Cattle raising began in northwestern Australia in the 1880s, with the overlanding of big mobs all the way from Queensland to the East Kimberleys.

Cattle are more climatically tolerant than sheep, and are more prominent in the hotter, drier, and more humid areas of northern, central, and coastal Australia. Due to a rising demand in both export and domestic markets, beef cattle numbers in Australia have almost doubled in the last quarter century. There are about 20 million head of beef cattle in the country.

The extensive raising of beef cattle is the dominant land use throughout the monsoonal north, in much of central Australia, and in most of eastern Queensland. In addition, more intensive breeding and fattening of beef cattle is carried on significantly in many parts of eastern. Queensland and northeastern New South Wales. Statewide distribution shows that nearly half of all beef cattle are found in Queensland and nearly one-fourth are in New South Wales, with considerably smaller totals in the other states and the Northern Territory.

Intensive breeding and fattening areas are mostly in hilly tracts on the coastal slopes of southeastern Australia. The cattle are often run in conjunction with some other type of farm operation, such as fat lamb raising or crop growing. The pastures normally are "improved"; that is, the woodland or bush has been cleared and exotic grasses have been planted. This permits a denser pattern of stocking and a much higher rate of turnoff of marketable animals. Most of the beef that is produced in the more intensive areas is sold on the domestic market.

The extensive cattle areas, on the other hand, are oriented toward the export market, and most abattoirs are in coastal towns from which the chilled or frozen meat can be shipped directly overseas. Many of the stations, especially in the far north, do no fattening at all and simply function as breeding areas. However, some cattle stations in the extensive area also provide some fattening facilities, in the form of either improved pastures or supplementary feeds, so that the animals only must be "topped off" for a brief period on better feed before slaughtering.

Cattle are usually mustered twice a year, once for branding the new crop of calves and once for selecting the animals to be sold. The rest of the year they fend for themselves. Shorthorns and Brahmans are the favorite breeds, and a cross between the two is commonplace (see Photo 4-5). Brahman cross cattle are now dominant in northern and central Australia.

The properties in the extensive area are, for the most part, quite large, and, although the legendary 12,000-square-mile (31,000 km^2) stations of the past have now been subdivided, there still are several that encompass more than 5,000 square miles (12,950 km^2) each.[5] Most cattle stations, then, require a number of full-time employees to handle the many chores associated with such vast enterprises. Aboriginal jackeroos (cowboys) used to be the mainstay of the extensive cattle industry, but in recent years they have been outnumbered by white employees.

Other Types of Livestock. There are some 2,500,000 *dairy cattle* in Australia, mostly found in the high rainfall country of the east coast. More than one-third are in Victoria, about one-fourth each in New South Wales and Queensland, and the other three states each have about 5% of the total. The optimum environment for dairying (cool, moist climate with a long growing season) is about the same as that for fat lamb raising, so these two activities are frequently found in the same areas. The principal dairy breed is the Australian Illawarra Shorthorn (AIS), which was developed in Australia and officially recognized as a distinctive breed in 1929; Ayrshires, Holsteins, Guernseys, and Jerseys are also common. Most dairy production is consumed locally, but a significant share is sent as milk or cream to a dairy factory for conversion into butter, cheese, or processed milk for export. The dairy

[5] For comparative purposes, the area of the state of Connecticut is just over 5,000 square miles (13,000 km^2).

Photo 4-5 *Shorthorn cattle drinking at a dry-season billabong along the Georgina River near Camooweal in northwestern Queensland. (TLM photo.)*

industry has been facing a price-cost squeeze of considerable magnitude, however, and is heavily subsidized by the government.

The number of *horses* in Australia is declining, as it is in most countries, but there are still about 400,000. They are useful in the extensive cattle-raising areas, so most are found in Queensland and New South Wales. There is also considerable breeding of race horses in Australia; horse racing is perhaps the most popular sport in a very sports-conscious nation.

There are some 2,500,000 *swine* in Australia. Most are kept as an adjunct to general farming, but there are some specialized pig raising operations. Their numbers are increasing steadily.

Poultry are widespread, but mostly raised on a small scale. There are some 50 million chickens in Australia, and perhaps another million of all other species of poultry. Since the late 1970s, there has been a notable upsurge of interest in *goat-raising*. A widening world demand for Mohair and Cashmere stimulated the rapid growth of Angora and Cashmere goat-keeping. Goats are also raised for their milk and their usefulness in brush and weed control. In the early 1990s, there were about 350,000 domestic goats in Australia.

Irreverence for the Land

This brief survey of the primary industries has shown that some of the natural resources of Australia are bountifully bestowed, whereas others are much more limited in quantity, quality, or both. Despite this uneven endowment, the pattern of resource exploitation has been one of short-run expediency, with little thought to tomorrow's consequences. This is an expected result when a relatively small number of immigrants invade a new and undeveloped continent; it has been characteristic of frontier settlements throughout human history.

In many ways, Australia is still a frontier country. The population is relatively small, and the land to be "conquered" is vast; thus frontier attitudes toward the land still prevail, although 20th century techniques are brought to bear. The bulldozer becomes the deity, and the cloud of dust becomes the symbol of contemporary Australian civilization. "Development" of the land is important—some would call it essential; but development based upon exploitation without conservation is rapacious. The sorry spectacle of 19th century resource depletion in North America has been and is being repeated in Australia. In spite of the prominence of rural industries in Australian consciousness, there has been little reverence for the land.

Until very recently, a positive approach to the land resource had only been manifested in three general endeavors—impounding surface waters, tapping artesian aquifers, and enhancing soil productivity with fertilizers. Most aspects of the environment have been subjected to mistreatment, including such environmental treasures as the Great Barrier Reef and delicate rainforests. On a broader scale, the practice of "ringbarking" (girdling the bark of a tree so that it dies) is surprisingly widespread in a land of few trees. The idea is to replace the trees with grass so that there will be more pasturage for livestock. More incomprehensible, perhaps, is the massive overgrazing that is characteristic of most pastoral areas. Animal husbandry is the dominant rural industry of the nation, yet millions of acres have been denuded and eroded because of overstocking with sheep and cattle. Graziers often are quick to blame drought, kangaroos, and rabbits for the problem, but objective studies have shown that the critical factor usually is overgrazing by livestock.

Probably the most serious environmental issue facing Australia today is land degradation. After two centuries of European settlement, land clearing, deforestation, and overgrazing, much of the Australian landscape is so destabilized as to be subject to a considerably degree of erosion and/or salinization. It is clear that much of Australia's agricultural prosperity in the past has been at the expense of the nation's soil capital.

Within the last two decades, however, a conservation consciousness of considerable breadth has arisen. Scattered adherents of the conservation crusade ("greenies" in the Aussie idiom) finally came together to stop the damming of a wild river (the Franklin) in southwestern Tasmania in 1983. This was the first greenie "victory" in the conservation versus development arena, and signalled the coming of age of the conservation movement. Beginning with a few dedicated individuals and expanding to organized groups, the citizenry has led the governments into more systematic and thoughtful consideration of resources/environment/ecology.

In the last few years, there has been an enormous surge of legislative activity with an environment-centered orientation that reflects policies of conservation and sustainable development. By the early 1990s, the notions of conservation, preservation, and sustainable development were encapsulated in the broad concept of "landcare," which was widely and rapidly being approved by the general public.

CHAPTER FIVE

Contemporary Australia

Most non-Australians have a mental image of the continent Down Under as a land of vast, dusty plains inhabited by hordes of sheep and kangaroos and peopled by hard-working, weatherbeaten country folk. This stereotype is enhanced by much of the literature of Australiana, which pays tribute to the "sunburnt land" and the hardy pioneers who occupy it. Such an image has validity in an areal sense, for most of Australia is both semiarid and rural, and the four-legged inhabitants are much more numerous than the two-legged ones. However, with respect to the Australians themselves, the stereotype is patently false. The people are overwhelmingly urbanites, the society is one of the most affluent in the world, the way of life is sophisticated and modern, and the economy is highly industrialized.

Australia's Urban Milieu

The typical Australian breadwinner lives in a suburban section of a large city, works five or five and one-half days per week in a factory, office building, or shop, and commutes to work in his/her own automobile. Her/his pattern of life is very much like that of the typical American. Thus, it comes as no surprise that 86% of the Australian population is classified as urban, whereas the corresponding figure for the United States is 76%.[1] Urbanism is clearly a dominant feature of the social geography of Australia.

[1]These two statistics are only roughly comparable, for in Australia towns with more than 1,000 population are classed as urban places, whereas in the United States an "urban place" must have a population of at least 2,500.

Centralization. Even more striking is the concentration of the urban population in a handful of large cities. The six state capital cities contain 58% of the continent's population within their metropolitan areas, whereas the approximately 400 smaller cities and towns include some 28% of the citizenry. In no other significant nation is the centralization of population so marked.[2]

Table 5-1 illustrates the primacy of the capital city in each of the states. In four of the six states the majority of the state population resides in the metropolitan capital, ranging from a high of 73% in South Australia to a low of 40% in Tasmania.

Such capital city dominance has been pronounced in Australia since the earliest days of settlement. We saw in Chapter 3 that the settlement of each of the Australian colonies was initiated at a coastal location that was sooner or later proclaimed the seat of government for the colony. These settlement centers, then, had the early dual advantage of being both principal port and administrative headquarters for their respective colonies. This led directly to their development as leading commercial centers as well, with the fundamental benefit of an early start. In logical fashion, the land transportation patterns of the various colonies radiated outward from the capital focus. The growing population provided an increasingly large market and labor supply, which attracted secondary industries to the capital cities. Thus almost the full range of economic advantages was concentrated in the metropolitan centers, and the agglomerative tendencies continued through the decades. In New South Wales, Victoria, South Australia, and Western Australia the capital cities are located in the approximate centers of the settled areas; their position reinforced their significance as primate cities of these states. In Queensland and Tasmania the metropolitan cities are off center, making it easier for other cities to grow to relative importance, and in these two states the capitals contain a significantly smaller proportion of both population and urban economic activities.

The Urban Hierarchy. The urban hierarchy is clearly dominated by the capital cities (see Table 5-2). Sydney and Melbourne are metropolises ranking among the 40 largest cities in the world. The four other state capitals (Brisbane, Adelaide, Perth, and Hobart) are of much less international repute and national stature, but they demonstrate notable primacy in their respective states. Hobart is only twice as large as Launceston in Tasmania, but in the other states the capital city is strikingly larger than any other urban area. The ratio ranges from 8:1 in New South Wales (Sydney: Newcastle) to 25:1 in South Australia (Adelaide: Whyalla).

Medium-sized cities are scarce (see Figure 5-1). Below the level of the state capitals, there are only six cities with populations exceeding 100,000

[2]For comparison, the six largest metropolitan areas of the United States contain 16% of the national population. Analogous statistics include Canada, 35%; France, 23%; United Kingdom, 19%; Japan, 16%; Brazil, 15%; and Germany, 12%. For New Zealand, the figure is 51%.

Table 5-1. *Capital City Population Dominance (1991)*

State	*Capital*	*Proportion of State Population*
New South Wales	Sydney	63%
Victoria	Melbourne	70%
Queensland	Brisbane	46%
South Australia	Adelaide	73%
Western Australia	Perth	72%
Tasmania	Hobart	40%
Northern Territory	Darwin*	48%

*Administrative center

Figure 5-1 Principal Urban Centers

(see Table 5-2). Smaller cities are somewhat more numerous; there are 15 in the size range between 40,000 and 90,000; most of which function as important regional commercial centers. Another 16 towns are in the 20,000 to 40,000 class, and more than three dozen have between 10,000 and 20,000 inhabitants.

In the states of New South Wales, Victoria, South Australia, and Western Australia a "typical" Australian urban hierarchy has developed. The capital dominates the life of the state; metropolitan population centralization ranges from 63% to 73% of the total. Secondary cities are much smaller, and within various size categories they are scattered more or less equidistantly from the capital, apart from a few specialized mining or smelting towns. Small towns are also dispersed concentrically inland from the capital.

In Queensland and Tasmania the patterns are more exceptional. The former state has an extensive coastline along which there are discontinuous patches of productive agricultural land extending for a thousand miles (1,600 km) northward from the capital. Furthermore, the rural productivity of inland Queensland does not decrease uniformly with distance from the sea as is the tendency in the other mainland states. Hence, there is an irregular pattern of rural population and a decidedly unusual transportation network. Long railway lines extend inland from four ports situated at regular intervals along the coast, rather than being funneled toward the capital. Brisbane, then, contains only about two-fifths of Queensland's population, and other

Table 5-2. *Australia's Largest Urban Centers*

Rank	*Urban Center*	*Approximate 1991 Population*
1.	Sydney, N.S.W.	3,500,000
2.	Melbourne, Vic.	3,000,000
3.	Brisbane, Qld.	1,200,000
4.	Perth, W.A.	1,100,000
5.	Adelaide, S.A.	1,100,000
6.	Newcastle, N.S.W.	450,000
7.	Canberra, A.C.T.	300,000
8.	Wollongong, N.S.W.	250,000
9.	Gold Coast, Qld.	230,000
10.	Hobart, Tas.	190,000
11.	Geelong, Vic.	160,000
12.	Townsville, Qld.	120,000
13.	Launceston, Tas.	90,000
14.	Ballarat, Vic.	80,000
15.	Toowoomba, Qld.	80,000
16.	Darwin, N.T.	80,000
17.	Ipswich, Qld.	70,000
18.	Rockhampton, Qld.	70,000
19.	Penrith, N.S.W.	60,000
20.	Bendigo, Vic.	60,000

cities have a combined share that is almost as large. Truly autonomous regional urban nuclei have evolved more pronouncedly in Queensland than in any other state, due to the more scattered distribution of productive land and other resources, and the "corner" location of the capital.

The pattern in Tasmania is somewhat analogous to that in Queensland, but on a much smaller scale. Hobart, too, is off center, and it is distant from the fertile agricultural and pastoral lands of the north coast and the mining areas of the western mountains. Thus Hobart's share of the state's population is actually less than the cumulative total of the other Tasmanian urban areas. Furthermore, the proportion of rural population is higher in Tasmania than in any other state.

Principal Cities

New South Wales. Australia's largest metropolis and most cosmopolitan city is *Sydney*. The initial Antipodean settlement of 1788 was on the shores of Sydney Cove, a small indentation on the south side of Port Jackson Bay, which is only a stone's throw from the commercial core of present-day Sydney. The deep and sheltered waters of the bay were superb for ocean-oriented commercial purposes, but they are not as suitable for accommodating contemporary urban sprawl. There are a great many minor coves and other indentations that fragment the rolling site of the city and have given rise to a maze-like street pattern. Further, there is only one bridge across the long bay, and, despite the highly efficient ferry system, urban development on the south side of the bay is much more expansive than it is to the north. Sydney is the leading commercial city, and vies with Melbourne as the largest industrial center of the country. Nearly one-third of all Australian manufacturing is found there; the industrial structure is highly diversified, reflecting the national structure.

Newcastle and *Wollongong* are the state's secondary cities of note. Both are primarily centers of heavy industry, combining the advantages of coastal location and nearby coal fields. Most of Australia's iron and steel is produced in these two cities, and there is also great emphasis on metal fabrication and machinery production. Cities of third rank in New South Wales include *Penrith*, and industrial and market center at the foot of the Blue Mountains west of Sydney; *Cessnock*, a coal mining town in the lower Hunter Valley; *City of Blue Mountains*, a chain of resorts along the main transport route west of Sydney; *Broken Hill*, the world famous mining town situated in splendid isolation in the arid west of the state; and *Albury*, a government-designated growth center on the border of Victoria that is situated on the direct highway and railway route between Sydney and Melbourne.

Victoria. Australia's other world-class city is *Melbourne*. It was founded near the head of Port Phillip Bay as a nucleus to serve a pastoral hinterland. However, the Victorian gold rushes of the 1850s worked major changes on Melbourne. Prosperity in the hinterland (based on both mineral and agricultural productivity) combined with strict protective tariffs for local manufacturing to engender rapid urban growth. During the last half of the 19th century Melbourne was larger than Sydney, and for the first 26 years of this century it was the temporary capital of the nation, until Canberra was finally sited. The economy of Melbourne is also broadly varied, and in the last few years it has exceeded Sydney in factory output. The industrial structure is well diversified, but there is particular emphasis on machinery, automotive, clothing, and shoe production.

Geelong, Australia's 11th largest city, is located on the edge of Port Phillip Bay some 50 miles (80 km) southwest of Melbourne. It is an important industrial city, characterized by large factories. Third echelon cities in Victoria are *Ballarat* and *Bendigo*, each located about 75 miles (120 km) (west and northwest, respectively) from Melbourne. Both experienced major gold rushes, and both have persisted as important regional commercial centers.

Queensland. The capital city of *Brisbane* has grown rapidly in recent years to become Australia's third largest urban center. The city was founded on the navigable Brisbane River some 10 miles (16 km) inland from the coast. Brisbane was chosen as the capital of the new state of Queensland in 1859 and has been the leading city in the state ever since. The city occupies a rolling site on both sides of the river, spreading eastward to the sea and westward into the foothills of the Great Dividing Range. Its function is largely commercial, although it has notable industrial production in food processing, wood products, and certain types of machinery. The subtropical location of Brisbane invests it with a character rather apart from that of the other Australian metropolises.

As noted earlier, Queensland contains several smaller cities of considerable regional significance. *Townsville* and *Rockhampton* are the principal centers of the northern and central portions, respectively, of the state; they occupy similar situations in that both have fertile farmland nearby and both serve productive pastoral and mining hinterlands. *Toowoomba* and *Ipswich* are inland cities, the former functioning as commercial center of the fertile Darling Downs agricultural region, and the latter a long-established coal mining and farming center in many ways analogous to Cessnock in New South Wales. Located in the extreme southwestern corner of the state, *Gold Coast* is a rapidly growing seaside resort agglomeration with a national clientele. *Cairns*, on the northeastern coast, is an internationally known tropical base for exploring the Great Barrier Reef, nearby rainforests, and the wilderness of the Cape York Peninsula.

South Australia. The focus of the planned colony of South Australia was *Adelaide*. It was laid out in a regular rectangular pattern on the narrow coastal plain between the Gulf of St. Vincent and the Mount Lofty Ranges. Its chief function has always been commercial, but it has also made a singular contribution to Australian industrial production, initially as a center for the farm machinery industry, and more recently in the output of automobiles, fabricated metals, and diversified machinery.

There are no second- or third-rank cities of significance in South Australia. *Elizabeth* is a government-founded industrial and residential center that dates from the mid-1950s; it was designed as a decentralized nucleus away from the capital city, but its location only 15 miles (24 km) from the commercial core of Adelaide predestined its absorption as a metropolitan suburb.

Western Australia. The capital of Australia's western state is in many ways the most isolated city in the world, some 1,400 air miles (2240 km) from its nearest urban neighbor. *Perth* is located on a broad reach of the Swan River some 10 miles (16 km) inland from the coast. It was the nucleus of a partially planned agricultural and pastoral colony whose prosperity received a significant boost with important gold rushes in the 1890s. Its economic structure is broadly diversified, as would be expected from its location.

Perth was the most rapidly growing capital city in the 1980s, and now has surpassed Adelaide as the nation's fourth most populous metropolitan area. Perth's outport is *Fremantle*, which has extensive docking facilities in the mouth of the Swan River; just south of Fremantle on the coast is the heavy industrial center of *Kwinana*.

The second largest urban area in the state is *Kalgoorlie*, the old mining town which has become the commercial center of the eastern goldfields; its population, however, is less than 30,000.

Tasmania. Hobart is the smallest of the state capitals. It was founded in 1804 as a supplementary penal colony on the southwest side of the broad estuary of the Derwent River. The city occupies gently sloping land between the river and the rocky sides of Mt. Wellington. Despite early prosperity, Hobart has generally experienced slow growth in comparison to the other capitals. It is a major apple shipping port, and it has a specialized industrial structure that emphasizes nonferrous metal smelting, wood products, and paper products.

The hub of Tasmania's north coast is *Launceston*, the largest of the third echelon cities in Australia. It is a port located at the upper end of the deep estuary of the Tamar River, some forty miles (64 km) from the open sea. It is the principal Tasmanian port for Bass Strait shipping and is noted for the production of machinery and woolen textiles.

Northern Territory. Although more than 60% of the population of the Northern Territory is classed as urban, there are only a few towns, and they are all small. The administrative center and only general port of the Territory is *Darwin*. It is the principal trading town of the "Top End" and boasts an impressive jet-age airport, but the economy of Darwin is based mostly upon government payrolls and other expenditures. *Alice Springs*, situated close to the exact center of the continent, is an important regional trading town and tourist hub, despite its small size and remote location.

Australian Capital Territory. The confederation agreement of 1901 stated that the federal capital should be located in New South Wales, but at least 100 miles (160 km) from Sydney. After a seven-year search, the location and boundaries of the ACT were decided upon, and a six-year international competition for planning the capital was held. Construction began in the 1920s, and in 1927 the capital was officially moved from Melbourne to *Canberra*. It is a uniquely designed city whose planning and construction have been alternately criticized ("a good sheep station spoiled") and admired from many angles. In recent years it has been the fastest growing city in the nation, and much of the criticism has been replaced by praise as the federal capital (like the Snowy River Scheme) has become a sort of national status symbol.

The Economy: Sophisticated, Industrialized, Paradoxical

The Australian economic structure has all the characteristics of a modern, industrialized nation. Primary activities employ only 7% of the work force, with rural industries (agriculture, pastoralism) accounting for most of that total. Secondary activities (manufacturing) give employment to 16% of the labor force. Tertiary activities provide the remaining 77% of total employment, with the principal categories being services, 25%; trade, 20%; and finance, 10%. Table 5-3 demonstrates the close correspondence between this structure and that of the United States.

Even without a consideration of population distribution, then, it would be clear from these figures that the Australian economy is strongly urban-oriented. A high proportion of manufacturing, services, trade, construction, and government employment would logically be concentrated in cities and towns, as well as a somewhat smaller proportion of transportation-communications-utilities employment.

Export Dynamics Until fairly recently, the Australian balance of payments was heavily dependent upon exports from the rural industries. Less than two decades ago raw primary products (such as wool, wheat, and ores) comprised nearly two-thirds of all exports, and processed primary products (such as meat, sugar, and canned fruits) made up another one-fourth. Today,

Table 5-3. *Australian and U.S. Economic Structures Compared (1990)*

	Proportion of Work Force	
Major Economic Activities	*Australia*	*United States*
Agriculture, Pastoralism	5%	3%
Mining	1	1
Construction	7	7
Manufacturing	16	19
Transport, Communications	7	7
Trade	20	21
Finance, Insurance, etc.	10	7
Services	25	31
Government	5	5
Other	4	4

however, manufactured goods comprise about half of all exports, with raw and processed primary products making up the other half. Thus, the growth of manufactured exports has been one of the most significant developments in the Australian economy since World War II.

Australia as an Industrial Nation. We have treated the rural industries in Chapter 4. It is worthwhile here to consider in some detail the manufacturing sector of the economy.

Secondary industry developed in Australia under several handicaps, the most important of which was isolation. During most of its formative history Australia had to depend upon overseas factories, particularly in Britain, to supply the great majority of needed manufactured goods. This dependence upon a distant, and therefore relatively expensive, source has persisted to the present with regard to a continually decreasing but nevertheless significant variety of items. However, with the emergence of many types of domestic manufacturing, often a slow and tortuous process under the aegis of major protective tariffs, the basic industrial dependence upon foreign sources has taken a different turn. As Australian manufacturers become established, the limitations of the domestic market set relatively narrow margins for expansion. Thus, isolation is reflected as a market problem rather than a supply problem. The cultivation and exploitation of overseas markets by the Australian manufacturer are difficult and limit opportunities for achieving economies of scale in production.

Protectionism played an important role in the historical development of manufacturing in Australia and is still significant for a few industries. Much of the early industrial lead in Victoria, for example, can be attributed in large part to a tariff policy that fostered the shoe and garment industries of Melbourne. After federation, the customs barriers between states were removed but were largely replaced by national restrictive tariffs. Protectionism may have been essential in the early years, but it does not foster dynamism in a more mature industrial nation. Furthermore, it encourages, or at least leads to toleration of, monopolies. In recent years, most tariff barriers have been removed. Along with deregulation of banking and airline transportation, this has had the effect of exposing Australia to the realities of international economics as never before, and the economy is still reeling from the impact.

Monopoly, or tight oligopoly, is a significant characteristic of many industrial classes in Australia. In such basic industries as steel, cement, glass, paper, and sugar, production is entirely in the hands of a single, or of a very few, corporations. In some other types of manufacturing, there is a quasi-monopoly at the state, rather than national, level. It has already been pointed out that the six colonies developed independently of each other, and in many ways the six states still form relatively discrete economic entities, a pattern fostered by their population distribution patterns and transportation networks. As a result, many industrial firms find it expedient to establish a factory in each capital city to serve the respective state markets.

A matter of some concern to Australian economists, and of even greater import to Australian politicians, is the degree to which foreign capital and foreign corporate enterprise have entered the Australian economy. This phenomenon is not restricted only to the manufacturing sector, of course, but it is particularly marked there. It is estimated that more than one-third of all Australian manufacturing is foreign owned, often through subsidiaries established in Australia, and another 5 to 10% foreign controlled. Such considerable foreign investment in secondary industry is a mixed blessing, with the advantages of risk capital and technical competence and the disadvantage of possible economic colonialism. In the past, most overseas investment was from Britain; today the United States and Japan are the principal sources.

The industrial structure of Australia is generally well diversified, reflecting the fact that Australian manufacturers supply most of the wide range of demands of the domestic consumer market. Table 5-4 portrays the contemporary industrial structure, as measured by employment in manufacturing. The principal types of manufacturing include the following.

Machinery. Although some specialized machinery is still imported from North America or Europe, most domestic machinery needs are satisfied by Australian producers. Generally speaking, the machinery industry is concentrated in Sydney and Melbourne, although there is increasing output in such smaller centers as Newcastle, Wollongong, Adelaide, and Brisbane.

Transportation Equipment. The automotive industry is the most important portion of this class of manufacturing. Auto making had an early start in Adelaide, where much of the actual manufacture of motor bodies is still

Table 5-4. *Australian Industrial Structure (1990)*

Major Classes of Manufacturing	*Proportion of Total Manufactural Employment*
Machinery	12%
Transportation Equipment	12%
Fabricated Metals	9%
Primary Metals	8%
Electrical Equipment	6%
Printing and Publishing	5%
Chemical Products	5%
Clothing	5%
Food and Beverages	5%
Wood Products	5%
Stone, Clay, and Glass Products	4%
Textiles	3%
Furniture	3%
Paper Products	2%
Leather Products	2%
Other	14%

carried on. Melbourne is the overall leader in production, however, with Adelaide strongly in second place. There have been several recent plant closures in Sydney and Brisbane. The industry is mostly composed of subsidiaries of American and Japanese organizations, with the principal producers being General Motors-Holden and Ford Australia. Railway rolling stock and tramcar manufacture are mostly done in government owned factories in the capital cities, especially Sydney. Shipbuilding is accomplished mostly in private shipyards; the major shipbuilding center by far is the small South Australian port of Whyalla. Aircraft manufacturing is limited.

Primary Metals. There is a flourishing iron and steel industry that utilizes domestic materials, supplies almost all of the domestic demand, and is actively seeking to penetrate foreign markets. Major production centers are Newcastle and Wollongong. The principal nonferrous metal smelters and refineries are located either near the mines (as at Mt. Isa), or at logical tidewater sites (as at Pt. Pirie, S.A., where most of the Broken Hill ores are smelted), or near major electric power sites (as at Hobart).

Electrical Equipment. This is a growing industry that is technically quite efficient. Most of the major production is by companies with overseas (especially the United Kingdom and the United States) affiliations. Two-thirds of the output comes from factories in Sydney, and most of the rest is produced in Melbourne.

Clothing. The knitted clothing and hosiery industries are well established in Australia, supplying most of the domestic demand. Other types of apparel are manufactured in smaller quantities. More than half the domestic output is from Melbourne; much of the rest is from Sydney.

Food Processing. This diverse class of manufacturing is widely dispersed over the settled parts of the continent. The larger flour mills are located in the capital cities, but smaller ones are scattered throughout the wheat belt. The three dozen raw sugar mills are concentrated in cane growing areas of Queensland, whereas the major sugar refineries are in the capitals. Meat packing plants tend to be market oriented, although a number of export abattoirs are situated in Queensland ports to expedite the direct overseas shipping of beef products. Fruit and vegetable processing is mostly accomplished in the horticultural districts, as in River Murray towns or near the capital cities. Dairy factories and breweries are mostly market oriented, thus concentrated in the larger cities.

As indicated by the preceding discussion, most Australian factories are associated with the major population concentrations. More than 90% of all manufacturing is found within the crescent-shaped coastal zone between Rockhampton (Queensland) and Whyalla (South Australia), and the majority is agglomerated between Newcastle (New South Wales) and Geelong (Victoria). Indeed, secondary industry is even more concentrated in the metropolitan areas than is population; in the five mainland states the proportion of manufacturing located in the capitals ranges from a high of 85% in Adelaide to a low of 55% in Brisbane. Sydney and Melbourne are the largest industrial centers, of course, containing between them nearly three-fifths of the nation's factories. Secondary industrial nodes are much smaller. Only eight other cities have as much as 1% of national industrial employment, ranging downward from 7% in Adelaide through Brisbane, Newcastle, Perth, Wollongong, Geelong, and Hobart to 1% in Ballarat.

The Post-Industrial Economy. Although manufacturing now plays a prominent role in Australia's foreign trade, its relative position in the domestic economy in recent years has diminished, as is true with most industrialized nations. The peak year in the history of Australian manufacturing was 1973. Manufacturing employment decreased by about one-fourth in the following two decades, although value added by manufacturing diminished to a much lesser degree, indicating higher labor efficiency and greater mechanization. The manufacturing decline was felt throughout the country. It was proportionately greatest in New South Wales and Tasmania, and least in Queensland and Western Australia. On the other hand, there were very rapid rates of expansion in finance and related services, and in community services, with strong absolute growth in both wholesale and retail trade.

Transportation: Critical and Costly. The basic geography of Australia is such that transportation needs and transportation problems are of primary significance. The land is vast, the population is sparse and concentrated in a few nodes that are great distances apart, the people are both mobile and affluent, much of the resource wealth (particularly mineral) is inconveniently distant from the populated districts, the export economy has long been dependent upon bulky primary products, and there are essentially no navigable inland waterways. Given this set of conditions, it is easy to see why Australians have paid much attention to the development of transportation.

The governments, federal and state, have taken a leading role in transport development, for the costs have usually seemed too great to entice much private investment. For example, Australia's early railroads were privately owned, but all of them developed such financial difficulties that they succumbed either to bankruptcy or government takeover. A significant portion of each government's contemporary budget is for transportation items, and more than 7% of the Australian work force is employed in the transportation-communications-storage triad.

Roads. It is physically less difficult and less costly to build roads in Australia than it is in many countries because much of the land is level and forests are uncommon. Nevertheless, a sparse population scattered over a vast land area calls for an extensive roadway network but does not provide adequate financial backing. Thus, Australia has more than 500,000 miles (800,000 km) of roadway, but fewer than 20% of the rural roads are paved and less than half of the mileage is more than sporadically maintained. Even so, the roadway network in the settled parts of the southeast is fairly complete, and a high proportion of the mileage in the densely settled areas is paved, although only a tiny fraction of the mileage is more than two lanes wide.

Some nine million cars, trucks, and buses are licensed in Australia. Most of them are operated in the densely settled areas, putting a distinct strain on the roadway system of these localities. In contrast, the Outback roads are relatively undertraveled. Australia boasts the third highest per capita vehicle ownership of any nation, so there are a great many cars, trucks, and buses on the roads. Most of the traffic is in and around the cities, of course, and traffic jams are a way of life for metropolitan Aussies. Roadway transport moves more than three-fourths of all domestic freight, but accounts for less than half of all ton-miles hauled.

Railroads. Railway building got an early start in Australia, when the first short line opened in 1854 between Melbourne and its outport. For three decades railway expansion was desultory, but beginning in the 1880s the rail nets expanded rapidly. Because settlement grew separately in independent colonies, isolated and discrete railway systems developed. In each colony (except in Queensland and to a lesser extent Tasmania, where, as we have seen earlier in this chapter, atypical conditions prevailed) the railway net centered in the port-capital and was totally uncoordinated with the net in adjacent colonies. Even today only 10 interstate border crossings are effected by rail lines.

The contemporary railroad pattern of Australia consists of four state-owned systems; a federal system that operates the transcontinental line as well as virtually all the routes in South Australia, Tasmania, and the Northern Territory; two specialized commuter networks (Melbourne and Adelaide); and half a dozen privately owned, ore-carrying companies. The picture is further complicated by the use of different gauges (width between the rails) in different areas, another legacy of the separate colonial railway systems.

The role of railroads in Australia today is much the same as it was in the past, except that their relative significance has declined. Their principal function is still to funnel the primary products of a state to or through the capital city. As in most countries, the railways are best adapted to moving the large load on the long haul. The principal commodities carried are coal, other minerals, and agricultural produce. There is considerable competition with trucking lines for freight and with airlines for passengers.

In summary it can be said that the Australian railway systems are both efficient and significant. Nevertheless, it is fair to note that they operate at a financial loss. In most years, only Queensland shows an operating profit on railway operations, and even this profit evaporates if loan interests and other overhead are charged. In a normal year, the five government-owned railway systems of Australia show a cumulative net financial deficit of more than $500 million.

Water Transport. Inland waterway transportation has never been important in Australia. There was some steamboat traffic on the Murray-Darling system during the latter half of the 19th century, but railway

Focus Box: The Railway Gauge Problem

In the middle of the 19th century, the six young Australian colonies were wrestling with the problem of starting railway systems. Among myriad decisions to be made was what width of gauge to choose. As a general consideration, a narrower gauge results in less speed and less volume of traffic, but is also much less expensive to construct and equip. As the six colonies were separate, each would make and implement its own decision. However, as the most populous and prosperous colonies, the New South Wales and Victoria decisions would be watched with interest by the other four.

Great Britain had finally decided on "standard" gauge, 4 foot, 8.5 inch (1,435 mm), for its railways in the 1840s, and the Colonial Secretary recommended this gauge to the Australian colonies. However, the Chief Engineer of the Railways of New South Wales was an Irishman, and he insisted on wide gauge, 5 foot, 3 inch (1,600 mm), perhaps because the Irish railways used that width. Accordingly, New South Wales adopted wide gauge in 1852, prior to the beginning of construction. This information was communicated to the other colonies, with the result that Victoria and South Australia also decided upon wide gauge.

However, the following year New South Wales appointed a different Chief Engineer (an Englishman), who arranged for the original plans to be rescinded in favor of standard gauge. It was some time before this change became known in the other colonies, and they decided that it was too late to alter their designs, so they went ahead with wide gauge.

The other colonies—Queensland, Western Australia, and Tasmania—adopted narrow gauge, 3 foot, 6 inch (1,067 mm), because of its economy of cost. South Australia also initiated narrow gauge in the northern, sparsely populated portion of the colony.

Thus the stage was set for impossible coordination among the railway systems of the various colonies. No two adjacent colonies had the same gauge except Victoria and South Australia. When the colonies federated in 1901 and became states, there began a lengthy series of discussions concerning gauge standardization, but action was very slow. Not until 1962 was there a common gauge (standard) on the railway line connecting Australia's two largest cities, Sydney and Melbourne. Since then the transcontinental line from Sydney to Perth has also been standardized, but almost everywhere else the anachronism of the colonial inheritance of diverse railway gauges persists—a classic example of the *non*-diffusion of an innovation.

competition put an end to that. Even so, water transport arouses considerable political interest in the vigorous interstate debates on water management along the River Murray.

Coastwise traffic is of great significance to the economy. Although only about 5% of all domestic freight travels by coastal steamer, long hauls are involved, amounting to about half of all ton-miles carried. Bulk products are the principal commodities handled; coal and iron ore make up more than half of the total tonnage, and other minerals, such as petroleum products and bauxite, comprise most of the remainder. Some 250 vessels are licensed for interstate or intrastate commerce. All are of either Australian or New Zealand registry, as these have legalized preferential position. Nearly half of all coastwise commerce is handled through the ports of Wollongong, Newcastle, and Whyalla, all of which are involved in the reciprocal flow of steel making materials.

Air Transport. Australians travel more miles by air per capita than do people of any other country. This is a fairly straightforward result of widely separated cities, a high standard of living, and good flying weather.

The Commonwealth government set up a Civil Aviation Branch in 1921, with the initial goal of developing air service in Outback areas where other means of transport were lacking. Indeed, the first route was established between the remote towns of Geraldton and Derby in Western Australia. City-serving runs were not begun until the 1930s, although the nation is blanketed with scheduled routes today.

Until very recently, air transportation was tightly regulated in Australia, which resulted in generally good service, but little competition and high costs to the consumer. Significant government deregulation took place in 1990, which has (at least temporarily) generated broader service, intense competition, and much lower fares. The four-fold hierarchy of commercial aviation continues to exist, but its formerly rigid boundaries have been significantly blurred:

1. The prime international carrier is Qantas Airways, a government-owned line that evolved from a small Outback company (Queensland and Northern Territory Aerial Services, whose initials are incorporated into Qantas). They began operating in 1922 and claim to be the second oldest air transport company in the world. Until deregulation, Qantas was the only Australian international carrier, but now the second tier airlines have begun service to New Zealand and Melanesia.
2. Prior to deregulation, there were two domestic carriers with nationwide service. A third company is now in operation at this level, and one of the long-time domestic carriers (Australian Airlines) has merged with Qantas, so that Qantas is now a domestic, as well as an international, carrier.
3. About a a dozen smaller companies, most of them subsidiaries of the national carriers, operate primarily as feeder lines. A couple of these have been expanding their routes to a sub-national level.
4. Another dozen still smaller companies function mostly at a regional level, usually within a single state; they operate small planes over short routes, serving smaller urban centers.

In general, Australian civil aviation service is efficient, heavily used, and characterized by an enviable safety record.

The use of aircraft in the Outback is also of great significance. Every pastoral property of any size has an airstrip, if only to accommodate emergency Flying Doctor Service. Air transportation of personnel and supplies is widespread, however, and such activities as seeding, fertilizing, and vermin control are often done from aircraft.

CHAPTER SIX

Australian Regions

Although Australia has a greater degree of homogeneity than any other of the settled continents, there is sufficient diversity that any understanding of the geography of the country requires a consideration of major regional characteristics. As in any system of regionalization, it is possible to subdivide Australia into an almost infinite number of regions. However, for the level of generalization needed in a book such as this, it seems useful to postulate a system containing six regions (see Figure 6-1):

1. The southeastern fringe
2. The mediterranean corners
3. Tasmania
4. The northeastern fringe
5. The monsoonal north
6. The dry lands

The Southeastern Fringe

The "heartland" of Australia consists of a crescent of land that extends from southeastern Queensland along the coastal fringe of the continent into the southeastern corner of South Australia (see Figure 6-2). Its inner margin is roughly 200 miles (320 km) inland. This region contains the largest cities, the busiest ports, the most heavily used roads and railways, the principal industrial areas, most of the densely settled agricultural and pastoral districts, a great many coastal resorts, and major coal-mining areas. It is by no means uniformly an area of dense settlement and bustling activities, however. Large areas of rugged hill country and dense forest are included; indeed, some of

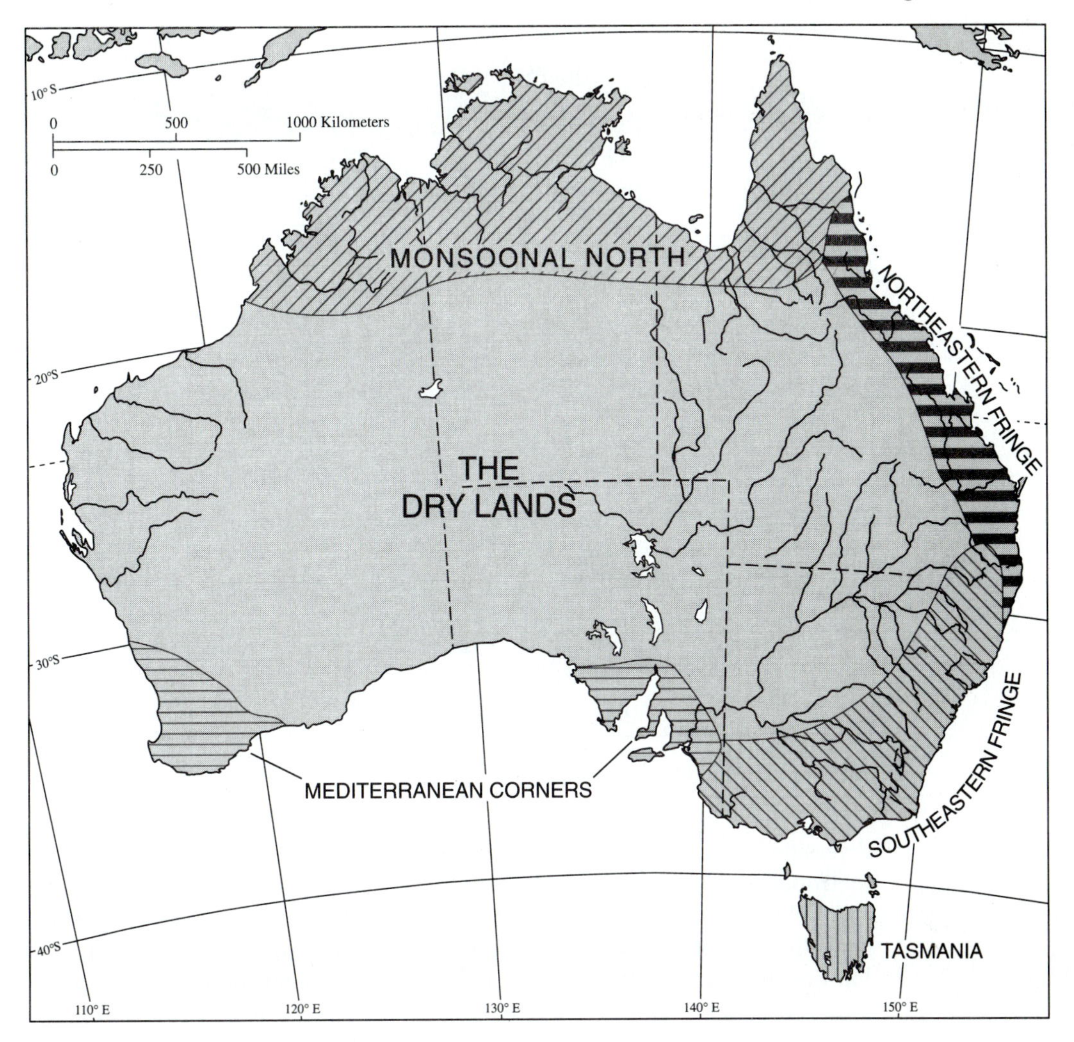

Figure 6-1 Major Geographic Regions

Australia's most impenetrable wildernesses are scattered through this coastal zone.

A generalized land-use model of the region would show a discontinuous series of small coastal lowlands and river valleys occupied by fairly intensive farming and pastoral enterprises and a number of variously sized urban agglomerations. These lowlands are backed by short but steep slopes leading into hill and low mountain lands that are partially covered with dense eucalyptus forests and partly grassed; these latter areas are devoted to grazing of sheep, cattle, and goats. Inland from the hill country is an expansive farming zone where sheep and grains are grown on properties of moderate size. This general pattern is interrupted in various places by irrigated districts of

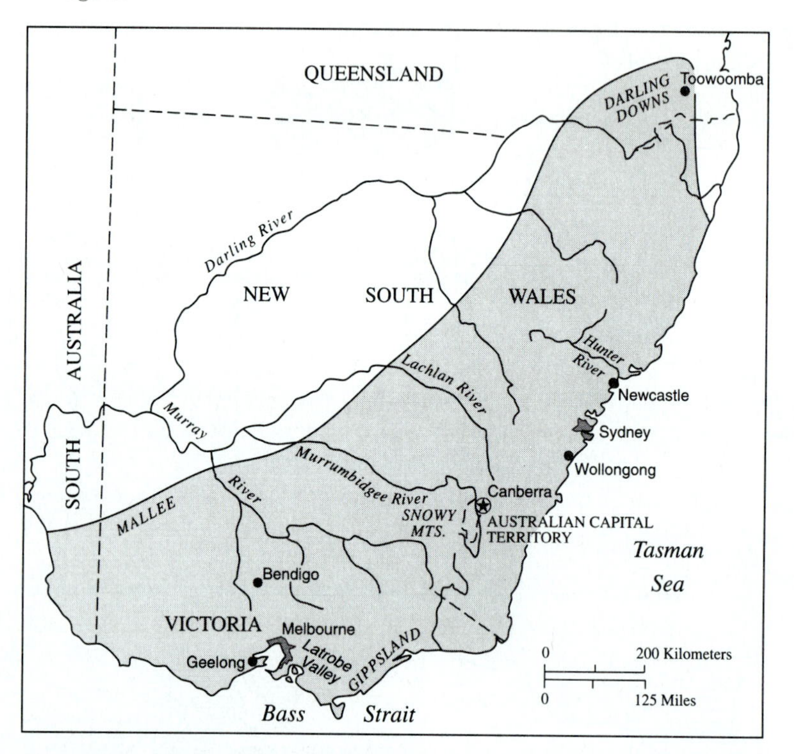

Figure 6-2 The Southeastern Fringe

diversified farm output, and the entire region is dominated by the major metropolises of Sydney and Melbourne.

This region contains most of Australia's population—some 60%. The concentration of people here is partly due to historical and political factors, but mostly relates to the more favorable environmental conditions, particularly the quantity and dependability of rainfall. Easterly and southeasterly winds, attracted by summer low pressures in the interior, bring Pacific moisture onshore. This moist airstream is forced up the slopes of the Eastern Highlands, which produces widespread orographic precipitation. In winter the region is dominated by westerly air flow that brings with it variable air masses, fronts, and storms. This, therefore is the only part of Australia, except Tasmania, that receives relatively reliable precipitation year-round, providing the basic necessity for dependable growing of crops and pastures.

Coastal Lowlands The coastal lowlands of this region are mostly devoted to farming and grazing activities, except where continually expanding urban settlements are preempting the land.

Farms in the coastal lowlands are usually of small to moderate size and are mostly family enterprises. Livestock products dominate, although there are many areas of specialty crop growing. The original vegetation of this zone was mostly forest and woodland. Where farming or pastoralism is carried on, the land has usually been cleared and replanted to exotic pasture grasses (such as subterranean clover or rye grass) or crops. The pastoral land normally is fertilized ("top-dressed" in Aussie parlance) with superphosphate, often spread from airplanes. This combination of clearing, replanting with exotic species, and top-dressing is generally referred to as "improving" the pasturage, and in most cases it greatly increases the capacity of the land to support grazing animals.

Dairying is the most widespread activity, with nearly two-thirds of Australia's 2.5 million dairy cattle found here. Most of the cattle are of European breeds, with the Holstein (called Friesian in Australia) and Jersey as leaders, although one "homegrown" breed, the Australian Illawarra Shorthorn (AIS, for short) is significant. The typical dairy farmer has a relatively small herd (40 to 60 animals), modern facilities, and a large debt at the local bank. The product of the individual farm is nearly always fluid milk, which is processed in the local creamery with about two-thirds of the output going into butter or cheese making.

The favorable environment for pasture growth is also an attraction for raising beef cattle. The improved pastures of the lowlands are capable of supporting about one beast per acre (2.5 head per hectare), and the typical farm of the area will carry from 2,000 to 3,000 head, of mostly Hereford or Shorthorn breed.

Sheep raising in this coastal zone emphasizes meat production, so dual-purpose breeds (such as Corriedale, Polwarth, and various crossbreeds) are more common than Merinos. A typical intensive sheep farm will contain 1,250 acres (500 ha) and carry from 15 to 20 sheep per acre (38 to 50 per ha). These are generally referred to as "fat lamb" enterprises, and the principal product is the meat of lambs slaughtered at an age of about 16 weeks. Coarse wool is a secondary product, and there is some slaughtering of older sheep for mutton.

Crop farming in the coastal lowlands is either a specialized enterprise (such as horticulture) or an adjunct to a livestock operation. Near the major urban centers the growing of vegetables and fruits for the local market is prominent; this is typical urban-serving market gardening, with small acreages and intensive farming techniques. In addition, there are a few areas of specialized farming that depend on broader markets; most notably the vineyards of the lower Hunter River Valley inland from Newcastle and the Gosford citrus area a few miles north of Sydney.

Mineral resources in this coastal zone are limited, except for major hydrocarbon—coal, petroleum, and natural gas—supplies. There are extensive measures of bituminous coal (called "black coal" in Australia) along the central coast of New South Wales. These coalfields are in a saucer-shaped configuration, roughly centered on Sydney. The seams lie several thousand feet beneath the city, however, and reach the surface in areas situated about 100 miles (160 km) to the north, south, and west. Much of New South Wales' electricity is generated by thermal plants on these coalfields, particularly in the lower Hunter Valley. These fields, particularly the southern ones inland from Wollongong, also provide most of the coal used in Australia's iron and steel industry. In addition, there is some export of coal to Japan.

In the Latrobe Valley, east of Melbourne, is the world's largest continuous deposit of lignite coal (called "brown coal" in Australia). Although the coal is low grade, it occurs in thick seams very near the surface, so it can be extracted inexpensively by open-cut methods. Latrobe Valley coal is used to generate most of the electricity for the state of Victoria.

Australia's principal economic mineral deficiency has always been petroleum. Oil is still the greatest single import to the country, but the domestic supply situation has improved considerably over the past quarter of a century. Most important by far are the underwater reserves of petroleum, and to a lesser extent natural gas, beneath Bass Strait off the coast of Gippsland (eastern Victoria). Six major fields have been developed, and their output now supplies about two-thirds of Australia's need for oil products.

The region, and the nation, are dominated by the two metropolises of the coastal lowlands. Sydney and Melbourne combined contain nearly 40% of the country's population; some 3.5 million in the former and 3.0 million in the latter. Although the sites of the two cities are quite different, their locational advantages and historical development are remarkably similar. Both had the benefit of being the initial settlement in a new colony; thus they got an early start as chief port, principal commercial center, and administrative headquarters. Soon the transport network of each colony was focused upon the port, and the funnel function between city and hinterland became deeply entrenched. As the colonies prospered, so did the cities. Labor, market, and capital were all concentrated in the two growing metropolises. The momentum of all these continuing advantages still persists, and despite much lip service and some concerted efforts at decentralization, the dominance of Sydney and Melbourne abides and even increases.

Sydney has an aesthetically splendid site around a deep estuary that provides a magnificent, sheltered harbor. The original community (a penal settlement in 1788) occupied the head of a little bay on the edge of a small coastal plain surrounded by rugged sandstone bush country. The center of Sydney's central business district evolved at this very spot (see Photo 6-1). If the settlement had an excellent site for ocean-oriented commercial purposes, it was lamentably unsuited to accommodate urban sprawl. Its several estuaries, with their many tributary bays, severely interfere with land transport, and the narrow twisting lanes of the early days only emphasize a contemporary street system that seems to have been designed by a madman. Sydney Harbour continues to be a beautiful bottleneck for land transportation, with only a single bridge for the first 14 miles (22 km) inland from the sea, although the opening of a cross-harbor tunnel, essentially paralleling the bridge, in 1992, has eased the congestion somewhat.

Urban sprawl is complete between Sydney Harbour and Botany Bay to the south; it spreads eastward into the Blue Mountains and southeastward into the Liverpool Plain. Sprawl is less extensive to the north, but, despite

Photo 6-1 *The central business district of Sydney, as seen from the north. The famous Sydney Harbour Bridge looms to the left. (TLM photo.)*

its uneven terrain, the land between Sydney Harbour and the Hawkesbury River is rapidly being over-run by urbanism.

Sydney contains the full range of urban functions that would be expected of a modern, primate city. It ranks either first or second to Melbourne in just about every conceivable Australian urban superlative (see Photo 6-2).

Major recent improvements include the laborious building of badly needed approaches to the Sydney Harbour Bridge, construction of a rapid transit rail line to serve the eastern suburbs, and redevelopment of the historic "Rocks" area near the Harbour Bridge into a major focal point of tourist interest. Significant projects now under way encompass the development of harbor facilities in Botany Bay, major expansion of the international airport, and the total reconstruction of Sydney's inner harbor (Darling Harbour) into a commercial/recreational zone.

Melbourne has a much less prepossessing site than Sydney, but is naturally much better situated to serve its hinterland. The city is located where a minor river (the Yarra) flows into the head of a major bay (Port Phillip). Fertile lowlands extend to east and west, whereas northward there

Photo 6-2 *Australia has a multitude of splendid beaches, and some of the finest are in Sydney itself. This is Bondi Beach on a mid-summer (December) Saturday. (TLM photo.)*

is a gentle access route (the Kilmore Gap) through the highlands to the vast River Murray plain beyond. Melbourne was founded in 1835, was chosen as capital of the colony of Victoria in 1851, and soon became the hub of transport. The Victorian gold rushes of the 1850s and 1860s greatly benefitted Melbourne, which served as the funnel to the goldfields in the same manner as San Francisco did to the California goldfields.

The land around Melbourne is flat to gently rolling, and there is little to interfere with the spread of an urbanized landscape except the broad lower reach of the Yarra River to the west of the city center. The West Gate Bridge, Australia's largest (even larger than the Sydney Harbour Bridge), provides the only crossing of the lower Yarra. Urban sprawl is very extensive to the north, east, and southeast, but less so to the west and southwest. The central business district was laid out in a rectangular grid pattern that emphasized wide streets, many of which originally were broad livestock trails, with ample avenues radiating outward in all directions to provide ready access to the countryside. As a result, traffic moves much more briskly than in Sydney, although not nearly as freely as one might expect.

Melbourne was Australia's largest city during much of the 19th century, although it has been relegated to a close second during the 20th century. Like Sydney, Melbourne is a dominant commercial, industrial, financial, and administrative center. Despite its bayside location, it lacked a good natural harbor. This deficiency was remedied by a threefold dredging and construction effort that has created adequate, if crowded, facilities near the city center along the Yarra, at the mouth of the Yarra, and directly south of the city center on the edge of Port Phillip Bay. There is considerable concentration of heavy manufacturing in these port districts, but other types of secondary industry are widely dispersed over the metropolis.

All of the Australian metropolises contain a considerable concentration of ethnic minorities, but none can rival Melbourne in this regard. Italian and Greek neighborhoods are particularly notable.

Forested Uplands. That portion of the Eastern Highlands that lies within New South Wales and Victoria is, for the most part, high and broad and rough enough to serve as a prominent barrier between the east coast and the vast plains of the interior. Much of it occurs in the form of a dissected plateau that is bounded on the eastern margin by abrupt escarpments, but merges gradually with plains on the interior side. In several localities the uplands extend right to the coastline.

Most of this upland country is clothed with a dense sclerophyll forest in which eucalypts are dominant. In some areas, it is a more open eucalyptus woodland. It is a zone of considerable precipitation, although variations in relief and exposure produce notable differences in the amount of moisture received within short distances. The highest elevations are in the Snowy Mountains section (where Mt. Kosciusko is located) of the Australian Alps of southern New South Wales, and continuing southwesterly into the Victorian Alps. The tree line occurs at about 5,900 feet (1,800 m), with alpine meadows forming extensive summer pastures above this level. This high country receives considerable snowfall for about three months, which supports a short but busy ski season at about a dozen localities.

Forest industries are not well developed except in the southeastern corner of New South Wales and easternmost Victoria, where there is considerable cutting of eucalypts for telegraph poles, railway ties, and lumber, as well as several large clear-cutting operations for pulp manufacture. In addition, a few large areas have been planted in conifers (mostly Monterey pine from California) to help alleviate Australia's severe shortage of native softwoods.

Some extensive tracts of land have been set aside as national parks or other forms of nature reserves. Most notable is the 1,540,000 acre (623,000 ha) Mt. Kosciusko National Park in New South Wales, although the Victorian government recently has reserved an immense tract of the Victorian Alps for national park status.

There is some livestock grazing carried out within the forests, but pastoralism is mostly found in areas of natural grassland or where the trees have been removed to foster grazing. Beef cattle generally occupy the wetter areas, including a distinctive transhumance development in the so-called "high plains" of the Victorian Alps. Sheep are much more numerous, however, primarily in the drier areas on the inland side of the ranges.

Crop farming is restricted to relatively small areas of fertile soil, often irrigated. Dairying is also prevalent in limited locales.

In a water-short continent it is to be expected that this relatively wet region would experience considerable "development" of its rivers, mostly in the form of dams. Both east-flowing and west-flowing streams have been dammed, the former mostly for urban water supply (especially in Sydney) and the latter primarily for downstream irrigation purposes. By far the most notable water development project is the Snowy Mountains Scheme between Canberra and Mt. Kosciusko, where two east-flowing rivers have been diverted via trans-mountain tunnels so that their water is used first for hydroelectricity generation and later for irrigation.

Although several medium-sized (20,000 to 40,000 people) market towns are scattered through the uplands, the only city of note is the national capital, *Canberra.* It is a completely planned city that was created in a pastoral valley in the southern highlands of New South Wales, the surrounding area being excised from the state to become the Australian Capital Territory (analogous to the District of Columbia in the United States). It is a beautiful city that has become the nation's urban showplace as well as the country's fastest growing city; both of these characteristics reflect generous federal funding.

Inland Farming Zone. Australia's fertile crescent is a 150-mile (250 km) wide swath of flat to gently undulating land that extends from the Darling Downs district of southeastern Queensland to the Mallee country of northwestern Victoria. It is generally the best farm land in the nation, with a combination of mostly fertile soils, adequate rainfall, a long growing season, and abundant sunshine. Numerous slow-moving rivers flow westerly from the uplands toward the interior lowlands, almost all of them eventually to become feeders of the River Murray or its major tributary, the Darling.

The natural vegetation was mostly savanna woodland—an extensive grassland with a scattering of trees, except along the streams where the trees grew more densely. Most of this has now been cleared to facilitate the growing of crops or pasture grasses.

Rainfall is normally adequate for grain production without irrigation; annual totals range from 15 to 35 inches (380 to 900 mm), with a summer maximum in Queensland phasing into a winter maximum in Victoria. The critical component, except in the Darling Downs where summer crops are grown, is the amount of winter rainfall because the principal crops (wheat, barley, oats) are all grown during that season.

The crescent has a considerable variety of soils, but includes some of Australia's most productive varieties. These range from rich blackearths in the Darling Downs to red-brown loams in New South Wales and lighter mallee sands in the west.

Focus Box: Trace Elements—Delicate Balance of Productive Soil

The productivity of any soil for plant growth depends upon its combination of physical and chemical characteristics. Most *mature soils* (those of reasonable depth and age) are likely to have an adequate balance of chemicals to nurture the "normal" natural vegetation of the area. When wild plants are removed, however, and replaced by crop plants or pasture grasses, there is often an inadequacy of chemical nutrients to support the new vegetation. In other words, when a farmer or pastoralist clears the land and attempts to grow crops or pastures, he is likely to find that there is a chemical imbalance in the soil that must be redressed before adequate yields can be expected.

There are four chemical elements—calcium, phosphorous, nitrogen, potassium—that are generally essential for productive agriculture. Calcium is critical to combat acidity. The other three often are referred to collectively as the *fertilizer elements*. Most commercial fertilizers contain large amounts of these elements; indeed, the N-P-K (nitrogen/phosphorous/potassium) ratio of the fertilizer is likely to be its most important selling point.

The soils of the Australian agricultural areas tend to be much more deficient in phosphorous resulting in applications of a phosphorous fertilizer element (usually in a form called *superphosphate*, or simply "super") by Australian farmers.

Less widespread but even more critical in some areas are deficiencies of *trace elements*. These are metallic salts that normally occur in the soil in minute quantities, but whose absence can significantly alter the productive capacity of the soil and/or the nature of the plants that grow in it. As agronomists learn more about the role of trace elements, they find that they can sometimes greatly enhance productivity by only small additions of one or more chemicals. These conditions exist in many parts of the world, but nowhere have they been demonstrated more dramatically than in Australia.

Many examples could be cited, but probably the most notable is from the southeastern part of South Australia, in an area formerly known as the 90-Mile Desert, and extending slightly across the state border into the so-called Big Desert of Victoria. A large Australian insurance company began an extensive land development program in this region in 1949, after obtaining large blocks of unimproved, "waste" land. The scrub was cleared, the land was plowed, and superphosphate and trace elements—mostly copper, zinc, and molybdenum—were added to the soil. Within 20 years, more than half a million acres (200,000 ha) were brought into agricultural productivity and the population of the area increased by 300%. Today, if you search the map of Australia, you will be unable to find "90-Mile Desert" or "Big Desert"; these areas have been renamed more in keeping with their present condition: Coonalpyn Downs and Telopea Downs. Local experts are in agreement that scrub clearance and superphosphate fertilizing contributed significantly to the development, but that the critical component was the addition of trace elements.

Similar success stories, on a somewhat smaller scale, could be told for the Esperance area of Western Australia, Kangaroo Island in South Australia, several districts in Queensland, and other localities across Australia. Thus, with the addition of very small quantities of metallic salts, it has been possible to bring literally millions of acres (an estimated 25,000,000 acres or 10,000,000 ha in Australia alone) of economically unproductive land into production, as well as enhancing yields from other expanses of land.

Early European settlement in the area was based on sheep raising, and Merino wool was the principal product during most of the 19th century. Grain, particularly wheat, was eventually introduced, and much of Australia's reputation as a major wheat grower was based on the produce of the crescent. Sheep again became important in the early years of the 20th century, giving rise to the wheat/sheep enterprises that still dominate the area.

The typical farm in this zone contains about 2,500 acres (1,000 ha), approximately one-fourth in grain cultivation (usually either wheat and oats, or barley and oats, with oats occupying a relatively small proportion of the acreage) and the remainder used for sheep grazing or the growing of fodder crops. It is all winter wheat, planted in autumn (April or May) and harvested in early summer (November or December). The harvested grain is trucked to a bulk storage facility, which is usually a cement silo. In years of bumper crops, the grain often is simply piled on the ground at a nearby railway terminal and covered with plastic sheets. The railways move the crop to major overseas terminals in the large cities or to flour mills for domestic processing. All marketing of Australian wheat is handled by a single authority, the Australian Wheat Board.

Typically, a farm will feed from 1,000 to 1,500 sheep, although the number may fluctuate significantly from year to year. If the farmer anticipates high wheat prices, he will plant more crop acreage and cut back on his sheep numbers. On the other hand, if wool prices are strong, he may plant less grain and devote more land to sheep grazing. Merinos are the favored sheep in this zone because wool is the principal product. However, many farmers concentrate on meat rather than wool, and use dual purpose breeds rather than Merinos.

The work necessary for wheat growing and sheep raising is complementary in that the busy times for one are not the busy times for the other. Thus a minimum of farm labor is necessary, except for sheep shearing which is done by contract shearing gangs. The average farmer can handle most of the other chores without the necessity of hired help.

This wheat/sheep pattern exists throughout most of the crescent except at the northern end (in the Darling Downs) and along a few irrigated river valleys. The Darling Downs has excellent soils and a preponderance of rain in summer rather than winter. The crop emphasis there is on such summer grains as sorghums and corn (maize), much of which is fed to or grazed on by beef cattle. Specialty crops are grown under irrigation in a variety of places, particularly along valleys and floodplains of west-flowing rivers (the Murray and its tributaries).

Irrigation is very prominent along the Goulburn River in Victoria and along the middle course of the Murray in New South Wales and Victoria, but the most notable district is the Murrumbidgee Irrigation Area (MIA), along the river of that name. Many field and orchard crops are grown in the MIA, and much land is devoted to irrigated pastures, but the most prominent output is a rice crop that boasts some of the world's highest per-acre yields.

The farm population of the crescent is spread fairly evenly, with a moderate density. As a result there is also a relatively uniform distribution of towns and small cities. The largest urban places are *Toowoomba* in the Darling Downs and *Bendigo* in central Victoria, each with more than 50,000 inhabitants.

The Mediterranean Corners

A distinctive climatic environment is found on the west coasts of all continents at about latitude 30° to 35°. This so-called "Mediterranean" climate is notable in that winter is the wet season and summer is virtually rainless. Australia has two southwest "corners"; one focused on Adelaide in South Australia and the other in the vicinity of Perth in Western Australia. These two areas are separated by the broad indentation of the Great Australian Bight, along whose coast the waters of the Southern Ocean meet the desert of interior Australia (see Figure 6-3).

The topographic patterns of the two "corners" are quite different. In Western Australia, there is a narrow coastal plain along the Indian Ocean that is separated by a low but abrupt escarpment (the Darling Scarp) from the dissected tableland of the great western plateau. In South Australia, an alternating sequence of three peninsulas and two gulfs has been produced predominantly by massive faulting. The land areas are characterized by north-south trending hills, mostly low.

The relatively rainy winters are occasioned by the "normal" weather patterns of the belt of midlatitude westerly winds. This airstream shifts northward in winter so that its storms and fronts pass over the Mediterranean

Figure 6-3 The Mediterranean Corners

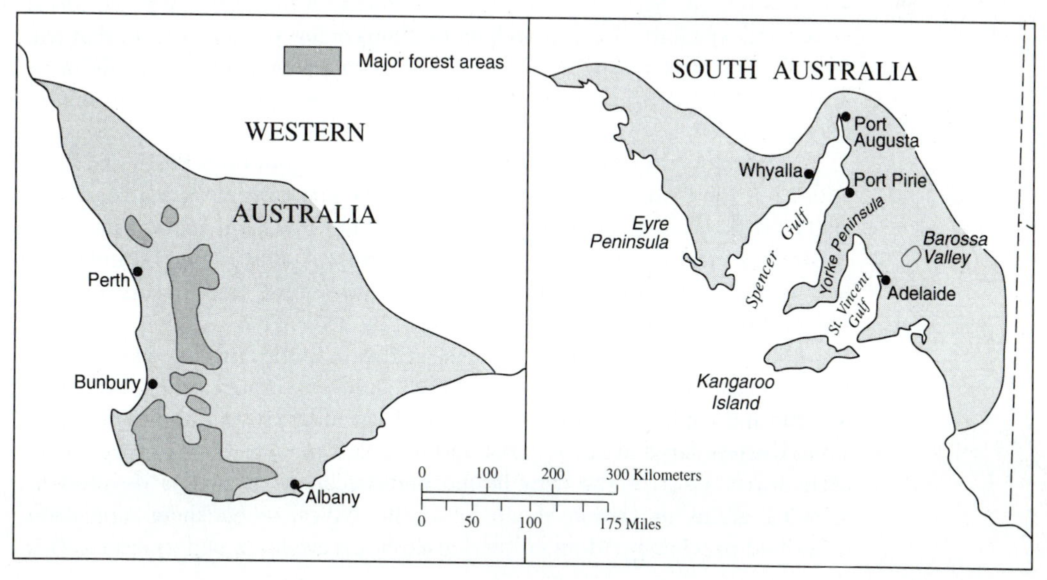

corners of the continent with some regularity. In summer, the wind and pressure belts are shifted southward, so that high pressure systems that produce hot, dry conditions control the atmosphere of the Adelaide and Perth areas.

Most of the world's Mediterranean lands have a natural vegetation cover that is dominated by shrubs. In Australia's mediterranean corners, however, forests and woodlands are extensive, presumably because many species of eucalyptus are quite drought resistant and thus able to cope with rainless summers. Forests are particularly notable south of Perth, where the wetter areas contain extensive tracts of gigantic *karri* trees (which can grow to heights exceeding 300 feet or 100 m) and the less humid zone has even more expansive forests of *jarrah* (which can reach 150 feet, or 50 m, in height). In the vicinity of Adelaide the forests and woodlands are considerably less vast, and the trees are less useful economically.

Exploitation of the karri and jarrah forests is a major activity in the southwestern corner of Australia. Both are strong, durable woods and the large size of the karri makes it particularly valuable. Most eucalypts produce inferior lumber, due to irregularities and blemishes in the wood, but both of these mediterranean timbers yield superior products. Much of the jarrah and karri country has been set aside in small national parks and large state forestry reserves.

Major management problems have developed in recent years, mostly associated with a disease called *jarrah die-back*, which is said to be the most destructive tree disease ever known in Australia. A fungus gets into the jarrah roots and inhibits the takeup of water and nutrients, slowly killing the tree. It attacks the jarrah, in particular, but it also affects other trees and shrubs in the district. There is no effective way to stop the fungus other than with intensive application of fungicide, which is quite impractical in a large forested region. Mitigation efforts focus on quarantining the infected forest areas, and concentrating research on methods of forest management that will help the forest to resist the fungus. Thus large sections of the jarrah forest are now off limits to all vehicles and all horseriders; they are open only to hikers.

Dairying and mixed farming are common around both Perth and Adelaide, but the distinctive agriculture of this region consists of irrigated specialty crops, as represented by orchards, vineyards, and market gardens. Australia's most famous wine area is the Barossa Valley, some 30 miles (48 km) north of Adelaide. But there are other outstanding vineyard areas in the hills south of Adelaide, along the course of the lower River Murray, and near the Darling Scarp east and southeast of Perth. Both citrus and deciduous fruits are also widely grown; particularly notable are oranges along the lower Murray and apples in several valleys about 100 miles (160 km) south of Perth.

Despite large areas of forest and notable concentrations of specialized agriculture, most of the area of the mediterranean region is devoted to growing grains or raising sheep or to the typical wheat/sheep operation described previously. Monocultural grain farms are more widespread in this region than anywhere else in Australia. Wheat is normally the dominant crop, except in parts of the mallee country southeast of Adelaide, where barley is the principal product. As wheat and barley are both winter crops, the lack of summer rain is no handicap.

The limited precipitation and dry summers, however, pose serious problems for the region as a whole. There is little or no perennial surface water. Even in the wetter areas, most streams stop flowing in summer, with the single exception of the River Murray, whose water source is far distant. Consequently, elaborate systems of water storage and transfer have been devised. In addition to many local reservoirs, the South Australian pattern is to move water from the lower Murray by lengthy pipelines to serve Adelaide as well as far-flung rural areas to the southeast, north, and northwest. In Western Australia there is considerable water storage on streams that flow westerly off the plateau just before they plunge down the Darling Scarp. Much of this water is used in Perth and in the coastal plain south of Perth, but lengthy pipelines also transfer water into the wheat belt and far beyond to the goldfields of the desert.

These mediterranean portions of Australia contain more than 90% of the people of their respective states, as well as being the only significant population concentrations not located close to the eastern or southeastern coasts. *Adelaide* and *Perth* are the capitals, chief ports, and dominant commercial, financial, and industrial centers of their states. Although different in terms of site and history, there is much similarity between the two cities.

Tasmania

The island state of Tasmania is separated from the mainland by the broad and choppy waters of Bass Strait (see Figure 6-4). Although the island is large, the state is small by Australian standards and encompasses less than 1% of the nation's area. Its population is also small—only about 450,000. Yet abundant rainfall provides conditions that have allowed a relatively high density of population, second only to Victoria among the states.

Probably the most salient element of distinction in the geography of Tasmania is its mid-latitude location, which sets it off from the rest of the country in terms of climate (colder winters, more dependable precipitation), natural vegetation (more trees, in general, and more and denser forests in particular), and water resources (an abundance of permanent streams). The popular image of Tasmania is a green island with apple orchards in the foreground and rugged mountains beyond. As with most stereotypes, this one is flawed but contains elements of truth. Green is certainly the prominent color for most of the island most of the year, but "drought" produces a typical brown Australian landscape for several weeks or months annually, at least in the eastern half of the state. Apples comprise Tasmania's principal crop, but orchards now are concentrated in only two small areas. However, there are mountains aplenty. Isolated peaks and discrete ranges are widespread, and

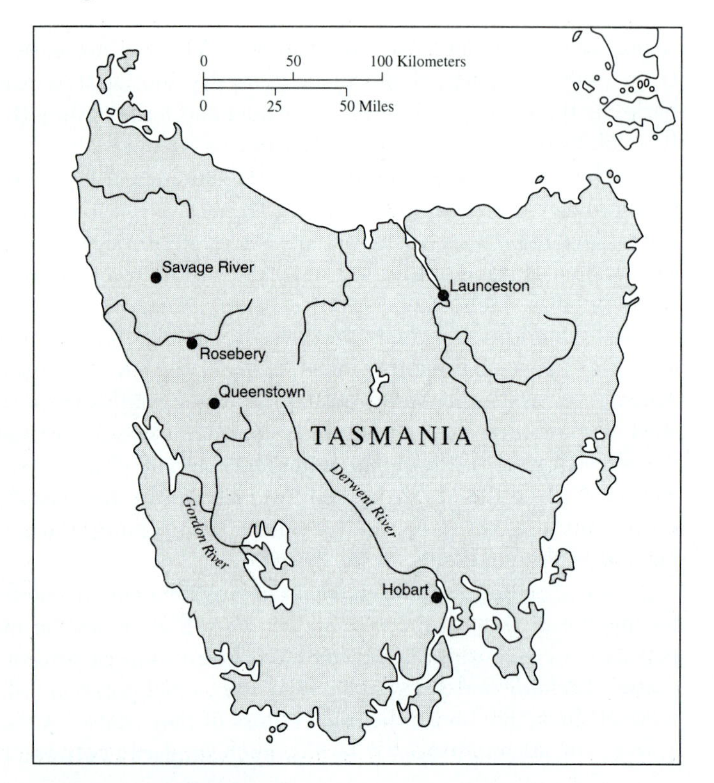

Figure 6-4 Tasmania

their high degree of dissection gives a more mountainous impression than most of the higher ranges on the mainland.

Topographically, Tasmania represents an extension of the Eastern Highlands across the relatively shallow waters of Bass Strait. Thus the terrain is almost entirely slopeland, varying from gentle to exceedingly rugged. More than one-third of the land is over 1,000 feet (300 m) in elevation, and a considerable area exceeds 4,000 feet (1200 m). Steep slopes, rocky outcrops, deep gorges, and a profusion of lakes emphasize the effect of glaciation during Pleistocene time, although no glaciers or permanent snowfields persist today. The center of the island is dominated by a massive plateau that is bounded on the north and east by an abrupt escarpment. Lowlands, virtually none of which are really flat, occur along the north coast, in an east-central corridor known as the Midlands and around Hobart in the south-east.

The mid-latitude position means that most weather systems approach from the west and arrive in winter; thus the westerly areas are the wetter ones and the winters see an abundance of precipitation. Tasmania is largely, but by no means entirely, forested. The wetter areas, including most of the southwestern quarter of the island, are clothed in a dense temperate rainforest (unlike anything else in Australia) although most of the species are eucalyptus. "Normal" eucalypt forests and woodlands cover most of the rest of the island, although there are areas of grassland scattered about, particularly in rain shadow situations or in the high country above about 4,000 feet (1,200 m).

Agricultural development has been limited by a lack of flat land and productive soils. Where favorable conditions prevail, however, there is mixed farming on a more intensive scale than commonly found on the mainland. Farmers tend to specialize in relatively high value crops, so that the quantity of output and its dollar value tend to be well above the Australian average.

Crop specialization capitalizes on the relatively cool temperate environment. Apples top the list, with intensive production coming from heavily laden trees (a notable Tasmanian landscape image shows small apple trees groaning under the burden of an enormous crop, with every limb propped up with timber to keep from breaking) in the areas south of Hobart and north of Launceston. Tasmania also produces the vast majority of Australia's hops and small fruits (raspberries, currants, loganberries, strawberries, and gooseberries), and large quantities of potatoes and peas are grown.

Yet the typical Tasmanian agricultural unit is a mixed farm, with a portion of the land given over to these specialized crops, but even larger areas devoted to more traditional items such as hay, pasture, and cereals. Farms are smaller than the Australian average, and there is an almost European emphasis on intensity and diversification of mixed farming. The resemblance is accentuated by the frequency of windbreaks and hedgerows in the agrarian landscape.

Livestock are also prominent in the farm scene. Sheep (primarily raised for meat) and beef cattle are widespread, but dairying is more typical. The striking black-and-white markings of Friesian (Holstein) cattle on emerald-green pastures comprise one of the enduring visual images of Tasmania.

The rugged western part of the state contains several major ore deposits. Major mining complexes at Queenstown (copper), Rosebery (lead, zinc, silver), and Savage River (iron) comprise the only significant areas of settlement/development in the western third of the island.

Tasmania's two most abundant resources—timber and water—are critical to the state's economy but are also focal points of conservation controversies that have become very prominent in the last decade. About one-third of the island has a forest cover that is economically useful. For decades, it has been supplying timber for lumber and for pulp and paper to the domestic market. More recently, woodchips have been steadily exported to Japan. The heavy and reliable precipitation of highland Tasmania nurtures numerous rivers that have been harnessed for the production of hydroelectricty. With 3% of the national population, Tasmania produces 10% of the nation's electricity—a per capita output unequalled elsewhere in the world except in Norway. Plentiful and relatively inexpensive electricity has attracted many power-hungry manufacturing operations to the state, particularly several involved in electrolytic refining of ores.

In the last few years, significant opposition to timber cutting and dam building has arisen. The latter particularly has been questioned because of the existing power surplus in the state and the high capital costs (mostly furnished by the state government) of further development. A plan to build a large hydroelectric dam on one of the last large free-flowing rivers (the Franklin) became a flashpoint of confrontation between "development" and "preservation" protagonists in the early 1980s. In the final analysis the federal government came down strongly on the side of the former, and the dam scheme was cancelled. This marked the first time in Australian history that a major development project was derailed for environmental-protection reasons, and was clearly a watershed in the evolution of conservation consciousness in the nation.

Tasmania has a much lower degree of metropolitan population concentration than any other state. Only about one-third of the populace lives in Hobart (see Photo 6-3). The second city, Launceston, is nearly half as large as the capital, which is unprecedented in other Australian states. The remainder of the population is spread fairly evenly over the agricultural districts, with a number of small cities along the north coast.

Photo 6-3 *Hobart, as seen from the north. The estuary of the Derwent River is in the foreground, and the forested slopes of Mt. Wellington rise in the distance. (TLM photo.)*

In sum, Tasmania is a region of scenic charm and economic disadvantage. Its "backwater" location is functionally remote from the Australian heartland, and its level of economic development reflects that fact.

The Northeastern Fringe

Most of the northern and central portions of Australia's east coast are included within a region that has a warm, wet climate (see Figure 6-5). This subtropical environment is favorable for the growth of a number of plants that are either absent or rare in the rest of the continent. Prior to European occupance, this coastal stretch contained many disconnected patches of tropical rainforest, which is unlike any other plant community in Australia. With the arrival of European settlers, tropical and subtropical crops—most notably sugar cane, bananas, and pineapple—were introduced, adding another note of distinctiveness to both the landscape and the economy. Offshore is yet another singular aspect of the region. It contains the world's largest coral reef complex with a superb and intricate ecosystem.

The region is long and narrow, nowhere extending more than about

Figure 6-5 The Northeastern Fringe

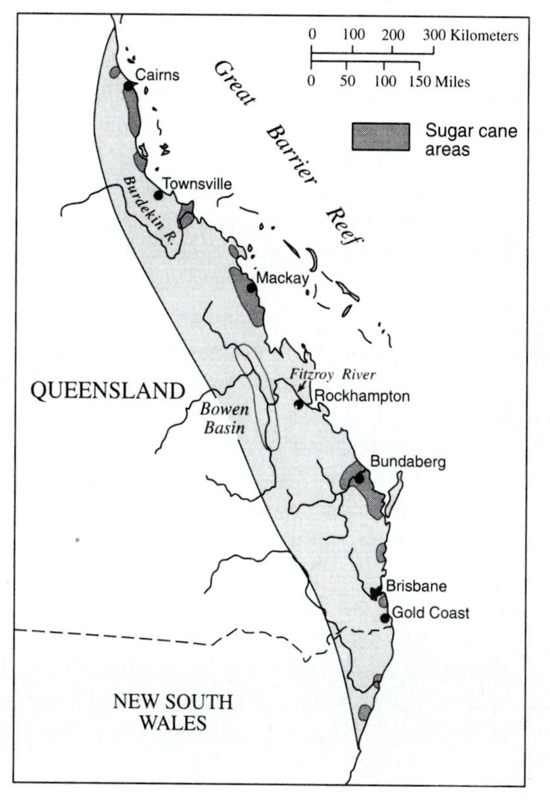

100 miles (160 km) inland because of the rapid diminution of rainfall away from the coastal slopes. Its topographic pattern is similar to that previously described for the southeastern coast—a mixture of small plains, river valleys, and low but steep-sided hills and mountains. Essentially, this represents the windward face of the Eastern Highlands, which descend steeply to the sea in some places but are interrupted by patches of lowland in many areas.

It is the climate that is responsible for much of the region's character. A tropical/subtropical location assures uniformly warm temperatures, except at higher elevations (which rarely exceed 4,000 feet or 1,200 m). Most of the region receives relatively heavy rainfall; generally more than 40 inches (1,000 mm) and, in some cases, well in excess of that figure. This is due to the persistent maritime breezes that are forced to rise up the topographic slope and drop the moisture they've accumulated from their traverse over the Pacific Ocean. Summer is clearly the wet season, but some rain occurs in every month.

Most of the region was originally tree covered. Eucalyptus forests were widespread, grading into a more open eucalyptus woodland in drier locations. Rainforest, however, was the distinctive component of the flora. This is a tropical plant association composed of species more typical of Indonesia and Malaysia; eucalyptus is almost completely absent from the Australian tropical rainforests. The trees grow tall and close together, with only limited undergrowth on the dark forest floor. Rainforest originally occurred in discontinuous tracts of varying sizes, always in locations favored by a frequency of persistent moist seabreezes; the largest expanses were in the southeastern corner of Queensland and in the far north of the region. Much of the rainforest area has by now been cleared for more "productive" land use, primarily dairying and crop growing.

European settlers came into this region from the south. A branch penal colony was established in 1825 where Brisbane now stands, but it only lasted for 15 years. Early pastoralists worked their way northward on the inland side of the ranges, making contact down the river valleys to the coast. Other settlers moved more slowly northward along the coast, seeking arable farmland. Gold rushes in interior Queensland, particularly in the 1860s, quickened the pace of settlement, as disappointed prospectors and miners turned to farming.

The early agrarian efforts were mixed farms supplying the small local market. Before long, however, attempts were made to grow specialized tropical crops. Several different crops were tried, and cotton looked like a winner during the American Civil War but could not compete after the war ended.

Sugar cane was first cultivated in the 1820s on a prison plantation at Port Macquarie on the north coast of New South Wales, but significant development dates from the 1860s in the Brisbane area. Originally cane growing was a plantation industry, depending largely upon *kanakas* (South Pacific islanders working under contract or indenture for a specified period) as a labor force. Sugar cane farming spread up and down the coast for 1,200 miles (2000 km), from Coffs Harbour in the south to Mossman in the north, leapfrogging from one coastal valley to the next in search of the right combination of soil, warmth, and moisture. Irrigation was introduced in several places to improve upon natural rainfall conditions.

There were many complaints and disputes about the *kanaka* system, with the result that their importation was prohibited in 1901 and almost all had been repatriated by 1906. This brought about a changeover from plantation cultivation with wage labor on large acreages to small, family-operated farms supplying company mills.

More than half of the 8,000 cane farmers in Australia are of Italian origin, and significant proportions are Maltese or Greeks. One of the myths of economic history is the assertion that Europeans cannot do hard manual labor in the tropics; the Queensland cane farmers have proved that this is not true.

The sugar cane farms stand in striking contrast to the eucalypt forest or scrubland that surrounds them. They are usually compact, unfenced blocks, with the farmstead encircled and virtually engulfed by the standing cane. At any one time on a given farm there will be checkerboard blocks of green growing cane of various heights, reddish plowed land that has been or soon will be planted, and perhaps a blackish standing crop that has just been burned. A typical farmstead will include a medium- to large-sized farmhouse and very few outbuildings. The farm house is usually built in the "Queensland style" (i.e., raised on high piles to allow free circulation of air beneath the house and also to provide a sheltered space for parking vehicles, storing equipment, hanging laundry, to deter vermin, and to provide a play space for children). Unlike houses in most of Australia, these homes usually have no chimneys, as fireplaces are uncommon.

The typical cane farm is about 70 acres (28 ha) in size. In any particular year, perhaps three-quarters of the land will be planted to cane, with the remainder in legumes or grass to recover fertility. There is usually a four-year rotation: the original planting year, two years of "ratoon" (volunteer) crops, and a resting year. Both planting and harvesting are now highly mechanized operations. The cane is usually set on fire the day before harvest in order to get rid of the trash and snakes in the field and to set the sugar content by driving off moisture. Burning makes harvesting an easier but dirtier job. One of the distinctive landscape features of this region during harvest time (June to December) is the cloud of black smoke rising from a burning cane field, with flocks of insectivorous birds (mostly small hawks called *kites*) wheeling above the smoke to snatch grasshoppers and other insects that are carried aloft by the convective updrafts of the flames.

After the cane fields have been mowed by gigantic mechanical harvesters (see Photo 6-4), the chopped-up cane is moved to the primary mills either by truck or via light railway cars (some 2,200 miles, or 3,500 km, of extra-narrow gauge line). The majority of the raw brown sugar that comes from

Focus Box: The Curse of the Cane Toad

The Northeastern Fringe Region is the adopted home of a chunky amphibian that was imported to increase sugar cane output, but has become a rapacious predator that imperils the ecology of a large part of the continent. In 1935, the South American cane toad (*Bufo marinus*) was introduced to serve as a natural control for two invertebrate pests that infested the sugar cane districts of Queensland. *Bufo marinus* is a large (the record specimen is 8.5 inches [22 cm] long and weighs nearly 3 pounds [1.35 kg]) toad with an insatiable appetite and prodigious reproductive capabilities—a female can produce 24,000 eggs per month. It has been introduced, with unpleasant results, into a number of Pacific islands (Hawaii, Tonga, Vanuatu, and Papua New Guinea) but its greatest impact, by far, has been in Australia.

For various reasons, the cane toad did not control the sugar cane pests for which it was intended. It has, however, become a major destroyer of virtually any animal its size or smaller. It has wiped out most of the native frogs, as well as a sizable share of the smaller reptiles, tiny marsupials, and terrestrial invertebrates, in its adopted habitat. Moreover, it is highly toxic to most carnivorous vertebrates, so any predator that eats the toad is likely to die as a result.

Most disturbing, perhaps, is the inexorable expansion of this adaptable creature's occupied range. It is now found throughout eastern and much of central Queensland, occupies the full extent of the Cape York Peninsula, has spread through the Gulf country into the Northern Territory where it reached Arnhem Land in 1992, and is moving southward into New South Wales. Presumably it will not be able to survive in cool or arid climates, but it has had a devastating effect on the local ecosystems in its areas of proliferation. *Bufo marinus* has the awesome potential of being the most destructive exotic animal to threaten Australia since the European rabbit.

these mills is exported to the world; the rest is taken by ship to large secondary mills in major Australian cities where it is refined into white sugar for domestic consumption.

At the time of confederation, in 1901, the cane growers of Queensland and northern New South Wales were guaranteed full supply of the Australian market. The growth of the sugar industry since then has been primarily through efforts to penetrate international markets. A high degree of efficiency in production and supervision enabled expansion to take place in a controlled fashion. Each of the three dozen primary mills is assigned a certain "crushing" quota each year, and that mill in turn makes acreage assignments to each of the 200 to 250 cane farms that supply it.

The Queensland and New South Wales Sugar Boards were moderately successful in developing expanded markets until the late 1950s when Cuba, previously the world's leading exporter, lost much of its overseas market. Australian production then mushroomed to fill part of this void and the sugar industry prospered and grew. Indeed, it eventually overexpanded, and the mid-1980s found economic disaster throughout Australia's sugar country, as the world price

Photo 6-4 *Mechanized sugar cane harvesting near Mossman in north Queensland. (TLM photo.)*

of this commodity plummeted and a majority of the Aussie cane farmers faced mounting debt payments they could not meet. Since then the sugar industry has been partially stabilized by improved government guarantees.

This region is also the principal Australian producer of other tropical crops. Most of the nation's bananas are grown on tiny farms (averaging 10 acres or 4 ha in size) situated on east-facing slopes along the north coast of New South Wales. Bananas could be grown much more widely in the region, and there has been recent rapid expansion in the northern part of the "sugar coast." But, for the most part, they are concentrated as far south as possible because that puts them closer to the major markets of urban Australia. Indeed, banana trains loaded with fresh produce set out daily from north coastal New South Wales for Sydney, Melbourne, and Adelaide. Most of Australia's pineapples are grown in the southeastern corner of Queensland, where they are processed in local canneries. The output of other tropical produce—papayas, mangoes, avocados, ginger, and various nuts—is scattered in this region.

Despite the prominence of specialized tropical crops, dairying is a much more widespread activity in these coastal areas. Nearly one-third of

Australia's dairy cattle are kept here (see Photo 6-5). Much of the pasturage is on slopeland that was originally forested; indeed, the clear-cutting of rainforest to be replaced by exotic pastures is one of the continuing conservation controversies of the region. Most of the output of the dairy farms goes into butter and cheese because the major fresh milk markets are distant.

A variety of other farming is mixed with the tropical specialties and dairy farms. An area of particular note is the Atherton Tableland, inland from Cairns in north Queensland. An elevation of 2,000 to 3,000 feet (600–900 m) reduces temperatures sufficiently to permit "unusual" crops at such a low latitude location. This is perhaps the foremost tropical dairying area in the world; it also has considerable output of tobacco, corn, and sorghums.

In the drier coastal stretches and on the inland side of the region, crop farming is rare and most of the land is either in forest or used for grazing livestock. This is a part of Australia where sheep proved unsuitable, and cattle breeding and fattening are the principal activities. These operations are moderately intensive, with a typical property containing perhaps 5,000 acres (2,000 ha) and 500 cattle. Much of the pasturage has been improved, and there is some supplemental feeding with hay and grain. Often cattle

Photo 6-5 *Dairy cattle (Jerseys) in the luxuriant hill country of southeastern Queensland. This scene is near Nambour, a few miles north of Brisbane. (TLM photo.)*

from outback stations in the interior are brought to this region to be fattened for a few months before slaughter in the abattoirs which are found in all of the larger Queensland ports.

On the western margin of this region in central Queensland is an outstanding series of coal measures in a structural depression called the *Bowen Basin.* A half dozen sizable mining complexes have been developed, with a new infrastructure of railway lines and ports to handle the commodity flow, which consists almost exclusively of direct exports to Japan.

Prominent rural industries notwithstanding, this coastal region is primarily famous throughout Australia for its subtropical coastal resorts, and known throughout the world for the wonders of the Great Barrier Reef. It is the premier tourist destination for urban Australians and is second only to Sydney as a goal of international visitors. There is a marvelous array of beaches throughout the length of the region; some are occasionally washed away in tropical storms, but they are always restored. Resorts, ranging from modest to super-luxurious, have been developed in dozens of localities, and many others are being built.

The Great Barrier Reef experience can be undertaken from several Queensland ports. Sometimes this is accomplished by day trips, but most "reefers" visit an island resort for one or more nights. Many of these resorts are not actually on the reef, but all of them provide a tropical seashore episode that serves the purpose.

The Great Barrier Reef represents a prominent area of environmental concern in contemporary Australia. Mineral exploitation, particularly for petroleum, comprises a potential "threat" to the integrity of the reef ecosystem, and "overdevelopment" is another arena of controversy. Large areas have been set aside as a marine national park and the entire reef complex has been declared a "world heritage area." But developmental conflicts continue. In addition, there is a notable natural threat to the reef—a relatively abrupt infestation of crown-of-thorns starfish. This voracious echinoderm is a devastating predator of living coral polyps and has ravaged extensive areas of the Reef, as it has reefs in other parts of the Pacific. The cause of this carnivorous eruption is still unknown, as is agreement on measures to deal with the problem.

International visitors flock to the Queensland coast in increasing numbers each year, but the economy of the region is stimulated much more by Australian tourists, primarily from the southern metropolises (Sydney, Melbourne, and Adelaide), who swarm to coastal Queensland (including the Reef) in winter as if it were the promised land. The principal destination by far is The Gold Coast, an attenuated resort complex in the southeastern corner of Queensland that was formed by the legal amalgamation of several previously separate urban places. The Gold Coast is referred to throughout Australia as "Surfers," an abbreviation of the name of one of the towns (Surfers Paradise) that was subsumed in the amalgamation. In form and structure The Gold Coast is similar to the Florida coastal segment from Miami Beach to Fort Lauderdale, although it lacks a large city as backdrop.

In 1986, it added to the scene an Atlantic City component—one of the world's largest gambling casinos.

The Gold Coast, with its dozens of hotels and apartment blocks, is the largest of the resort complexes, but there are many others. The Sunshine Coast—which begins about 30 miles (48 km) north of Brisbane—is clearly "in" with Australians of the 1990s in part because of lower prices. Another major focus is Cairns, at the northern end of the region. This eminently tropical city (latitude 17°) combines all of the tourist attributes of the region: fine beaches, a nearby reef island, easy access to the outer reef, excellent "big game" fishing, tropical agriculture, nearby rainforests, etc. Many other examples of favored coastal resorts could be cited; suffice to say that for most Australians the Queensland coast is their favorite wintering place.

Some 2.5 million people, about one-seventh of Australia's population, reside in this region. The distribution of the rural populace is a direct reflection of the land use pattern: forest areas are unpopulated, beef cattle districts are sparsely settled, areas of more intensive land use (sugar, dairying, etc.) have moderately high densities. Growing and prosperous towns and ports have evolved to serve the region, although the current crisis in the sugar industry casts a pall over all.

Most of the populace is urban. About 45% of the region's people live in greater Brisbane, which is the third largest Australian metropolitan area. As with the other capital cities, Brisbane dominates its state's economy and politics, but to a lesser extent than the other mainland capitals because of the impressive economic development of Queensland's coastal zone, which is mostly far removed from Brisbane.

Thus, other prominent third-level (as elsewhere in Australia, there really are no second-level cities) cities have evolved along the Queensland coast. *Townsville* (population 100,000), in the north, and *The Gold Coast* (250,000), in the south, are the largest. Other urban places of prominence are *Rockhampton, Cairns, Bundaberg,* and *Mackay.*

Perhaps more than any other part of Australia, Queensland represents a state that is "different." And with 90% of Queensland's population here, we can note that this region represents that difference. In many ways, Queensland is analogous with the Deep South of the United States; indeed, it has been referred to more than once as Australia's Deep North. This analogy is based largely on the "conservative" philosophy of many Queenslanders: capitalism, state rights over federalism, patriotism, love of family, suspicion of outsiders, distrust of labor unions, etc. As a reflection of these feelings, the state government was controlled for many years by a conservative administration that favored free enterprise, mistrusted Canberra, encouraged development, and was generally satisfied with the status quo.

Ironically, in recent years this region has become the principal growth center of the counterculture movement in Australia. Attracted by the mild climate and other pleasant environmental conditions, people seeking "alternative lifestyles" have flocked to this region. The result is an increasing frequency of political clashes, with significant economic and social overtones.

The Monsoonal North

Roughly the northernmost 20% of the continent—the area north of about 19° of latitude which encompasses large parts of Western Australia, the Northern Territory, and Queensland—can be identified as a region that is distinctive in character because of the dominance of a monsoonal climate (see Figure 6-6). All monsoonal systems function in approximately the same fashion, although the complexities of causality are still imperfectly understood. Apparently, the Australian monsoonal system is connected with that of Southeastern Asia in a complicated fashion. In simplest terms, low pressures develop over the warm Australian continent in summer, attracting a moist onshore airflow that produces a rainy season; during winter, high pressures occur over the land, occasioning offshore airflow that brings a dry season.

Thus, in northern Australia, during the December–March period, there are persistent winds from the north that bring warm, moist, unstable air. Parts of the *Top End* and of Queensland's Cape York Peninsula, experience more than 60 inches (1,500 mm) of rain in an average summer, but the amount of precipitation diminishes rapidly interiorward, so that along the 19th parallel of latitude the quantity is only about 22 inches (550 mm). Tropical cyclones also occur somewhere in the region two or three times each summer, on the average, adding to the rainfall total and sometimes causing serious damage. The dry season is much longer than the wet in northern Australia, generally lasting for about seven to eight months. The summer wet season is accompanied by temperatures that are monotonously warm to hot; the dry season has somewhat greater temperature fluctuation, varying from moderate to hot.

Figure 6-6 The Monsoonal North

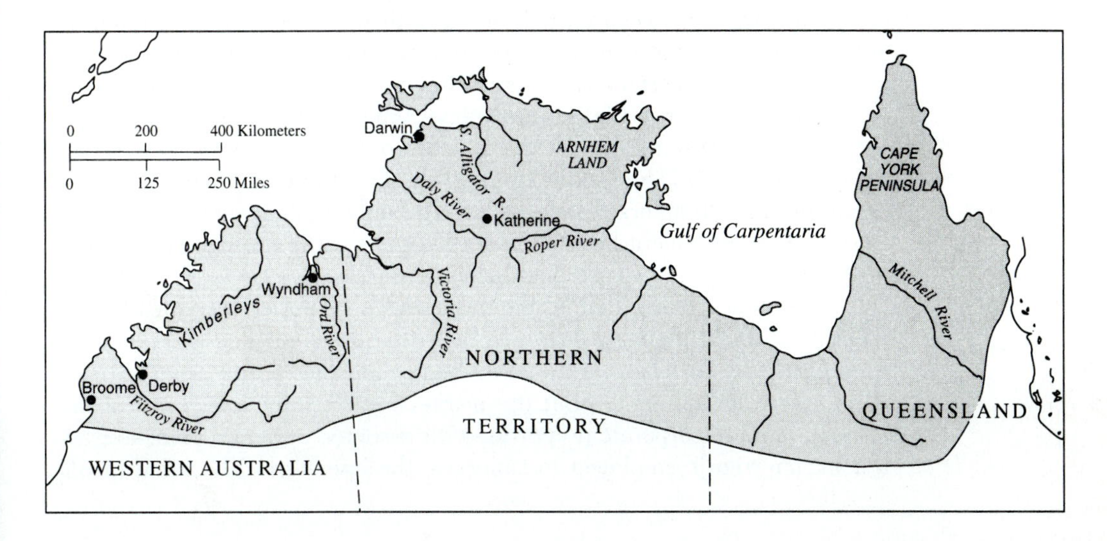

The natural vegetation that has evolved under this specialized climate is largely a grassy woodland. The tree cover consists mostly of various species of eucalyptus growing in an open pattern so that there is considerable space between trees, although there are a few small patches of dense forest, even rainforest. Almost everywhere a prominent grassy understory spreads beneath the trees. In summer the grasses grow green and tall, but as the dry season sets in they rapidly become brown, tangled, withered, and generally non-nutritious to herbivores. Moreover, they provide abundant fuel for the wildfires that rage throughout the region almost every "winter."

Soils of the monsoonal north are generally unproductive. The heavy summer rainfall erodes and leaches, producing a predominance of stony or sandy soil types. The most fertile soils are along river floodplains, but these areas are generally flooded and/or waterlogged during the summer growing season.

The majority of the region consists of extensive flattish plains, seamed by many rivers that may be raging torrents at times during summer but are often reduced to trickles or total dryness in winter. There are extensive rugged, rocky hills in the Kimberleys District of Western Australia and a prominent, scarp-edged, much dissected sandstone plateau in Arnhem Land.

Overall, the environmental conditions are harsh for human settlement and activities: too much rain for part of the year, too little for the rest; extensive flooding in the flat lands during summer; occasional destructive storms; generally infertile soils; unpalatable vegetation for much of the year; an abundance of insect pests; a variety of wildlife hazards (snakes, crocodiles, dingoes, box jellyfish, termites); a long distance removed from the heartland of the nation.

Aborigines occupied this region in somewhat greater densities than they did in most other parts of the continent, which is essentially a reflection of the relative abundance of food resources that could be collected on the land and in the adjacent seas.

The early European settlers were mostly either adventurers, visionaries, or both. The governments of the day failed three times in its efforts to establish a north coastal port before Palmerston (later to be re-named Darwin) was finally successfully founded in 1869. Many rural development schemes were planned, but all failed sooner or later, usually sooner. Sheep raising was a total failure; crop farming, an almost total failure; and mining was sometimes temporarily successful. The only abiding activity that supported rural settlement by whites was extensive cattle raising, and that has sustained only a very sparse population density.

The typical cattle *station* (the term "ranch" is not used in Australia) is a large leasehold property on which many hundreds or thousands of cattle fend for themselves most of the time, being "mustered" only once or twice a year for branding or to draft the marketable stock for sale. Most of the stations now are corporate properties with absentee owners represented by a manager who is employed to supervise the operation. The hired hands (*stockmen* or *jackeroos*) are sometimes Aborigines, sometimes white men, and increasingly white women (*jilleroos*). Much of the station work is still done on horseback, but there is increasing use of motorbikes and airplanes.

Most of the "turnoff" from monsoonal cattle stations is exported for the production of manufactured beef (McDonald's hamburgers, etc.). There are several export meatworks in the northern ports that operate during the dry season but close down during the arduous "wet."

An interesting outgrowth of the beef cattle industry in the Top End is the raising of domesticated water buffalo. The ancestors of these buffalo had been introduced as domesticated animals at the time of the early settlement attempts in the mid-19th century. When the settlements failed, the buffalo were turned out to shift for themselves, and found a happy home in the seasonally inundated floodplains of the Top End, establishing a viable feral population that numbered about a quarter of a million by the early 1980s. These "wild" buffalo had been hunted for hides and meat for many decades, but not until the last few years has any systematic effort been made to redomesticate them for raising as a ranching operation. There are now half a dozen stations in the area that concentrate on buffalo husbandry. Meanwhile, in a massive effort to combat actual and potential livestock diseases, primarily tuberculosis and brucellosis, the government had eliminated most of the feral buffalo by the early 1990s.

Apart from cattle and buffalo raising, the economic output of the region is limited largely to a handful of farms, a few large mines, and some specialized fisheries. An expensive and elaborate agricultural scheme on the lower Ord River in the northwestern corner of Western Australia finally began to make good in the late 1980s, a quarter of a century after its inception. The only other successful farming in the region consists of a variety of relatively small horticultural operations on fertile alluvial soils around Katherine, 220 miles (350 km) southeast of Darwin. Thriving mining enterprises have been developed laboriously and expensively at a half dozen localities. Commercial fishing is represented particularly by prawning in the Gulf of Carpentaria and a dozen cultured pearl enterprises in sheltered northern bays.

The monsoonal north is one of the few parts of Australia where Aborigines comprise a sizable proportion of the population. Many of them are fringe-dwellers in the towns, undereducated, underemployed, occupying the lowest rung of the socioeconomic ladder, and depending largely on government welfare payments for their livelihood. Others occupy traditional tribal areas, maintaining some tribal cohesion and cultural identity; this is most notable in Arnhem Land. Within the last decade the federal government has enacted legislation that pertains only to the Northern Territory (there is nothing similar in Western Australia or Queensland) whereby Aborigines may gain title (via collective Land Trusts and Land Councils) to "traditional" lands. Through this mechanism about one-third of the area of the Northern Territory portion of the monsoonal north region has come under Aboriginal control. As with all land in Australia, the government retains the mineral rights.

However, Aborigines have a veto power for exploration or mining ventures, with the result that Northern Territory Aborigines have received considerable royalty payments from mining companies, and many Aborigines are now employed in the mining operations.

The only urban place of any size in this region is the administrative center of the Northern Territory, *Darwin*. Although long established, Darwin has never had a viable economy without heavy infusions of government spending. The city was virtually demolished by a devastating cyclone that hit on Christmas morning, 1974, killing more than 60 people. The population dropped from 50,000 to 5,000 after the storm, but government-sponsored rebuilding has been spectacular. Darwin now has a population of more than 70,000 and is one of the fastest growing cities in the country. Apart from government spending, however, the only economic support is tourism. During the dry season Darwin is the transport hub for large numbers of tourists who come to experience the "true tropics" and to see the remarkable Aboriginal cave paintings and wildlife of Kakadu National Park, 135 miles (215 km) to the east.

In the more remote parts of Australia, there are three prominent threads of political/social/economic concern about the use of the land that are at least in partial conflict with one another:

1. "development" to provide economic benefit to an impoverished region;
2. preservation of the pristine environment; and
3. land "rights" for the Aborigines.

All three are conspicuous in the monsoonal north region and can be exemplified by the confrontation that took place over a potential mining development in the drainage of the South Alligator River on the eastern edge of Arnhem Land. The details are complicated, but in simplest terms the area in question was considered:

1. to be significant traditional land by the local Aborigines;
2. to have outstanding scenic and environmental virtues by conservationists; and
3. to have rich reserves of uranium ore.

Conservationists opposed any development; Aborigines opposed development without their full consultation and approval as owners of the land; and the Territory government and the business sector of the economy favored full development without delay. After several years of litigation, consultation, and financial lubrication, a compromise was reached in which some (but not all) of the ore deposits were opened to mining with strict environmental-protection regulations in effect, a large national park (Kakadu) was established, Aboriginal land ownership was affirmed, and the local Aboriginal councils were paid significant royalties for their cooperation. This solution left many people unsatisfied, but it at least signalled that rapprochement is possible in contentious land use matters, even among bitterly opposing factions.

The Dry Lands

Australia's largest, most remote, and least densely occupied region is a vast arid and semiarid expanse here called *The Dry Lands* (see Figure 6-7). This region comprises almost two-thirds of the area of the continent, but contains only about 6% of the continental population. It is too dry for farming without

Figure 6-7 The Dry Lands

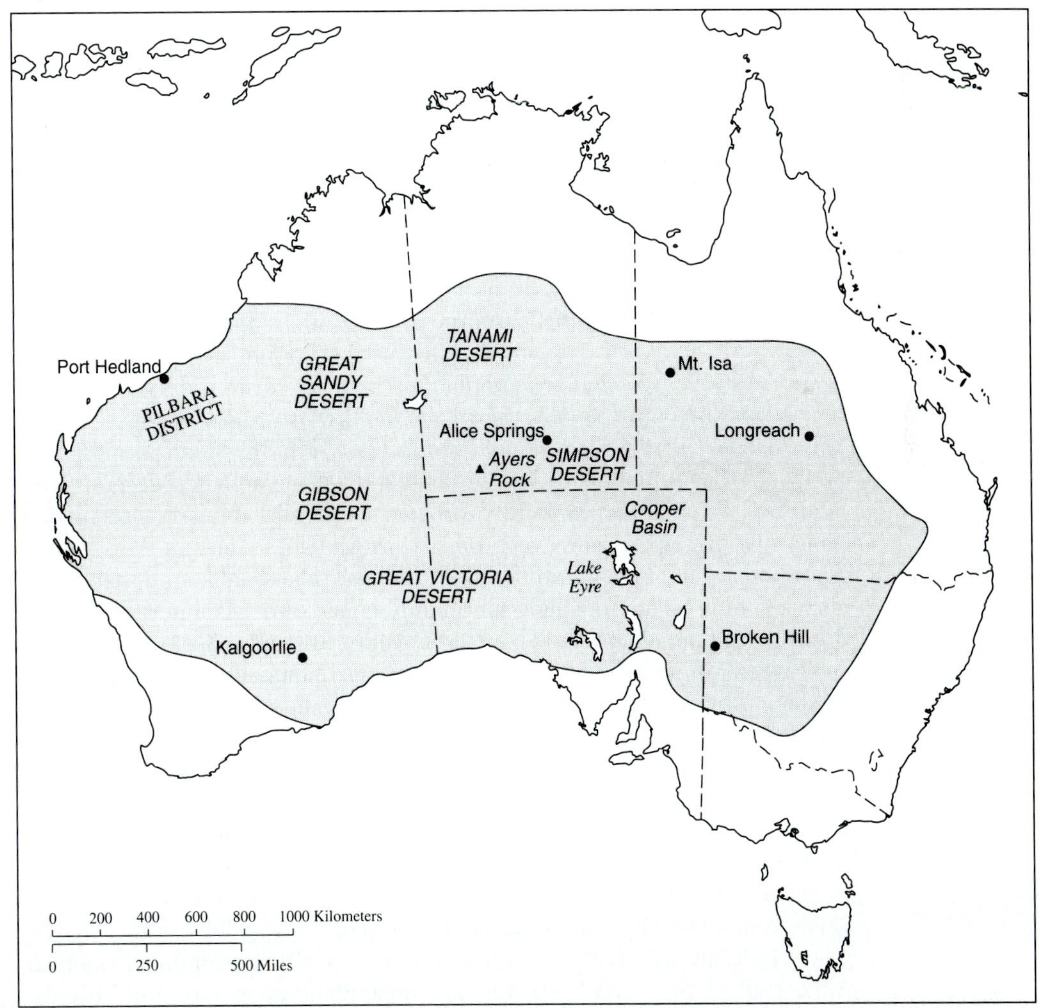

irrigation, too remote from water sources for irrigation, and too barren for animal husbandry other than extensive grazing. Thus human occupance is restricted mostly to widely dispersed pastoral homesteads, scattered Aboriginal settlements, and a limited number of small towns.

The geologic fundament of the Outback is so old as to be almost timeless. Ancient Precambrian rocks underlie much of the region, outcropping extensively at the surface in some places, but overlaid with a relatively thin veneer of younger sedimentaries or still more recent unconsolidated sands elsewhere. The landforms that have developed on this ancient surface are, for the most part, very subdued. In some places there are rocky monoliths or ranges that rise above the general level, but the fundamental topographic dimension is horizontal, not vertical. Some very extensive sections, epitomized by the Nullarbor Plain (where there is a 300-mile, or 480-km, stretch of the transcontinental railway that is perfectly straight and perfectly level) and the Barkly Tableland, have profiles that are as regular and featureless as can be found anywhere on Earth. The flatness is not absolute elsewhere, but horizontality dominates the lay of the land, and in only a few areas (mostly notably the Pilbara country of the northwestern shoulder of Western Australia and the MacDonnell Ranges of The Centre) is there a significant amount of slopeland and relief.

Lack of relief does not imply smoothness of surface, however. As in most arid parts of the world, even the minor topographic features tend to be barren, rocky, and rugged. Sparseness of vegetation and thinness of soil combine with a predominance of mechanical weathering to emphasize the starkness of bare rock, the abruptness of hillside slope, the steepness of canyon wall, and all the details of differential erosion.

Certain specialized topographic associations recur widely within the region, but are of infrequent occurrence in other parts of Australia. Particularly notable are the longitudinal dunefields which are found in most arid parts of the continent (completely dominating the five named deserts—the Simpson, Tanami, Great Sandy, Gibson, and Great Victoria), occupying thousands of square miles with their repetitious parallelism. Also widely dispersed are the sprawling beds of intermittent lakes, so often misleadingly shown in blue on maps, which represent the end points of numerous internal drainage systems that are totally without water the vast majority of the time. Less extensive in occurrence, but uniquely characteristic of the region, are the flat gibber plains, surfaced as far as the eye can see with little rocks and tiny stones.

Impressive, isolated eminences rise above the general landscape level in various places. They are usually rounded sandstone or granite projections of erosional origin with steep sides and long slopes. The most spectacular are found in the southwestern corner of the Northern Territory, in the form of gigantic monoliths: Ayers Rock is the most famous, but nearby Mount Conner and The Olgas are equally spectacular.

Hydrographic patterns within the region also are often distinctive. Almost all of the surface streams are dependent upon rare and episodic precipitation, so that their regime fluctuates between long periods of little or no flow and brief intervals of flooding. The Lake Eyre Basin in the eastern portion of the region is by far the largest basin of interior drainage in the country. Its extensive watershed experiences major floods about three times a century, with the floodwaters collecting in a vast inland sea that is actually below sea level. Aridity prevails, however, for the vast majority of the time, and the bed of Lake Eyre is usually a dusty plain. Over more than half of the region drainage networks are only minimally developed; indeed, the Australian Dry Lands contain the largest area in the world with uncoordinated drainage.

Despite a scarcity of surface waters, there is an unusual quantity of underground water beneath portions of the region, particularly in the northeast (Queensland and vicinity). This is the Great Artesian Basin, one of the most extensive reservoirs of underground water to be found anywhere in the world. The quantity of water in these aquifers is huge, but its usefulness is severely limited by three major factors—depth, heat, and particularly mineralization. In sum, it is a resource that can be tapped only by means of expensive "bores" (wells), and when the water reaches the surface, it is too hot and salty for any use other than watering stock. Nevertheless, several thousand bores have been sunk, mostly by the government, which has permitted a considerable expansion of sheep and cattle raising in central and western Queensland.

The central theme of the climate of The Dry Lands is the dominance of episodic drought. For most of the region drought is a continuum that is interrupted only sporadically by brief intervals of rainfall. The northern margin receives about 22 inches (560 mm) of precipitation, but most areas receive less than half that amount. Rainfall is not only scarce but also unreliable, and great variations from the norm can be expected in any given year. The clear skies and direct sunlight of summer produce very high temperatures; readings in excess of 100°F. (38°C) can be expected for many weeks. Night temperatures are considerably lower because the heat is quickly radiated away in the clear atmosphere. Winter days are usually warm to hot, but nights cool significantly, with most parts of the region experiencing below-freezing nocturnal temperatures, at least occasionally.

The natural vegetation of the region is a mosaic of considerable diversity—a variety of grasses and shrubs, as well as occasional clumps of trees. Sparseness of plant cover is to be expected in such a dry environment, but in actual fact there is a surprising amount of vegetative biomass. Grasses predominate in more northerly and easterly areas, whereas shrubs are more notable in the west and south. Trees occur mostly along watercourses. The most widespread plants in the region are the rounded hummocks of spiky grass called spinifex, which are particularly prominent in the sandridge deserts. Some areas are totally devoid of vegetation, but their extent is small, as compared to the whole. Australia contains nothing like the great "sand seas" of the Sahara.

The Dry Lands are the classic home of the Aborigine. Aboriginal people occupied every part of the continent prior to the coming of Europeans, but they were soon displaced from or inundated by whites in the more favored areas. The harsh environment of The Dry Lands was less attractive to the whites and offered fewer opportunities for them, so the Aborigines were less pressured, and have maintained a significant presence in the region.

Well over half of the total Aboriginal population of Australia is found in The Dry Lands Region. They live in a variety of situations: some are town fringe-dwellers, some are wage-earners on cattle stations, increasing numbers (but still a small proportion) are employed at mines or other enterprises, most are concentrated in distinctive Aboriginal settlements. Some of these settlements were founded and/or maintained as church missions for paternalistic ministering to Aborigines, most were set up by the government as Aboriginal "reserves" (similar to our Indian reservations), and some have been established by the Aborigines themselves (often with significant government funding) to serve as focal points for Aboriginal concentrations in a white society. Schooling and social services are generally available in these settlements, but gainful employment is scarce, and social problems (alcoholism, petty theft, assault-and-battery, drug use, boredom, etc.) are rampant.

Within the last decade, the Aborigines have expanded their political and economic power considerably, primarily as a result of land rights legislation enacted by the federal government (for the Northern Territory) and the state government of South Australia. Freehold title to vast acreages, including the central Australia area that encompasses Ayers Rock and The Olgas, has been granted to various Aboriginal Land Trusts and Land Councils.

It is worth noting that women often are the keystone to Aboriginal family and social life, but have virtually no influence in political affairs. Policy decisions are inevitably made by males. Various sociological studies have demonstrated that female opinion frequently is diametrically opposed, but this is an irrelevancy in Aboriginal society.

Apart from the Aboriginal presence mentioned above, most settlements in the region are pastoral stations, and by far the most pervasive land use is the extensive grazing of livestock. With the notable exception of some areas in Western Australia, there is a clear-cut separation of wool and beef operations. A pastoralist raises either sheep or cattle, but not both. Merino sheep dominate on the eastern and western margins of the region; it is no coincidence that these are the better-watered and better-grassed portions. Beef cattle (predominantly of either Hereford or Shorthorn breed) are found elsewhere.

The sharp demarcation between these two herbivorous species is often marked by the presence of a dingo fence, a 6 foot (2 m) high barrier of wire netting built for the specific purpose of keeping dingoes out of sheep country. Dingoes frequently attack sheep, but generally do not molest cattle; thus dingo country is cattle country, and considerable efforts are made to exclude them from sheep country.

The smallest of the pastoral properties are on the eastern margin of the region, where there is a relative abundance of nutritious native grasses (as well as underground water from the Great Artesian Basin) that can support roughly one sheep per acre (0.4 ha). Most other sheep areas in the region have a much lower carrying capacity—on the order of 25 to 100 acres (10 to 40 ha) per sheep. Cattle stations also are sparsely stocked, although in many cases even a light stocking is too heavy for the environment, and overgrazing persists as a significant problem.

Many of the pastoral properties are like remote, self-contained communities, able to exist with little or no external contact or support for long periods. Some of the largest of the stations have as many as 100 people living around the homestead.

Various schemes have been devised to "develop" some portion of The Dry Lands, but, with limited exceptions, only mining has been successful in this endeavor. Even mining is successful only for as long as the ore bodies hold out. Although smaller and/or briefer operations have been fairly widespread, in only five localities has mineral extraction been a persistently productive activity:

1. Broken Hill is a long-established (1884) silver/lead/zinc producer in western New South Wales that has become something of a transportation center and is actually the largest urban place in the region.
2. Mt. Isa in western Queensland is a more recently established enterprise (1923) that has the potential to become the world's largest single producer of lead, zinc, silver, and copper.
3. The Western Australia Goldfields district, centering on Kalgoorlie, was an outstanding gold producing area in the 1890s that has become prominent again for its output of gold and several base metals, most notably nickel.
4. In the last quarter century there has been an extraordinary development of iron ore mining in the Pilbara country of Western Australia. New towns, new ports, new railways, and a half dozen major open pit mines have been constructed in this wild and remote area, with most of the output being exported directly to Japan.
5. The Cooper Basin, in the northeastern corner of South Australia, has become Australia's principal source of natural gas, with lengthy pipelines taking the output to Adelaide and Sydney.

The lure of the desert is finally being exploited as a tourist attraction that entices more and more short-term visitors to the region. *Alice Springs*, situated almost in the exact center of the continent, is the hub of this activity, with new hotels, golf courses, and even a gambling casino in this unlikely spot. Three hundred miles (480 km) to the southwest by a newly paved highway is Uluru National Park, which contains Ayers Rock and the Olgas; the twin allure of "The Alice" and "The Rock" brings several hundred thousand visitors to the center of the continent each year.

Focus Box: Outback Airborne Amenities

Living in The Dry Lands tends to be harsh, drab, and lonely, particularly for women. Inhabitants of pastoral homesteads may be dozens of miles from the nearest neighbor, and the availability of such "ordinary" requisites of civilization as telephones, medical care, and schooling may be hard to come by or totally unavailable.

Innovative Australians, however, have confronted these problems with imagination, and have invented two institutions to make life much more tolerable in The Dry Lands. These have received world-wide acclaim and been copied in several other countries:

1. *Royal Flying Doctor Service.* In the mid-1920s, a Presbyterian minister, Rev. John Flynn, concerned about the inadequacy of health care in the Outback, conceived the concept of a flying medical service. His idea was made feasible by the invention of a simple, foot-operated, sending-and-receiving radio transceiver (the "pedal wireless") by Adelaide engineer Alf Traeger. From humble beginnings in western Queensland, the RFDS network has spread to encompass two-thirds of the continent.

Although designed primarily to serve pastoral homesteads, the system is functional for any remote settlement. Every homestead has a sending/receiving radio (a modern outgrowth of the pedal wireless) and a well-stocked medicine chest. In each of a dozen Outback towns, there is an RFDS base that includes a powerful radio station, an airport, and a hospital. A doctor attends the radio base for morning and afternoon sessions, during which he broadcasts medical advice to any client within reach of his signal (each of the dozen service areas encompasses an acreage larger than Texas). Moreover, an emergency attendant monitors the radio at all times, for any crisis that may develop. If the service of a doctor is needed, one will be flown to the location of the need (thus almost every Outback homestead or other settlement has an adjacent airstrip); if a patient needs an emergency trip to hospital, a Flying Ambulance is dispatched.

Each of the 12 services is separately administered, but all function in much the same fashion. They are supported partially by government funds, but significantly by charitable contributions.

As wireless telephone service is being extended widely into The Dry Lands in the 1980s and 1990s, the infrastructure of the RFDS may be significantly changed, but its basic function will continue.

2. *School of the Air.* In order to bridge the isolation and loneliness of Outback children, the concept of a radio school was envisaged shortly after World War II, and became operational in 1951. Operating mostly over the RFDS network, radio lessons are conducted between a teacher sitting in her/his empty classroom adjacent to the radio base and her dozen or so pupils scattered over thousands of square miles of remoteness. The school operates for elementary grades only, with each grade receiving 30 to 60 minutes of air time per day.

These periods of aural contact are buttressed by written lessons carried on by correspondence. In addition, all students come to their classroom once or twice a year to meet their teacher, meet their classmates, have a barbecue, and take exams.

Life in The Dry Lands is still difficult, but it has been ameliorated significantly by the addition of these two airborne institutions.

CHAPTER SEVEN

Australian Problems and Prospects

The area of Australia is almost exactly the same as the area of the 48 conterminous United States, but its population is only one-fifteenth as large. As we have noted, there are major environmental differences between the two countries. Moreover, despite basic demographic similarity, the patterns of population distribution and the ethnicity of the two countries are extremely divergent.

Nevertheless, for an American, Australia is fairly easy to comprehend because the common Anglo-Saxon background has produced a multitude of cultural/social/political/economic similarities. The American will soon note certain "peculiarities" in Aussie society—a parliamentary political system, voting requirements that are both mandatory and preferential, comparatively strong state governments vis-á-vis the federal government, a relatively high degree of government regulation of many aspects of everyday life, a powerful and militant complex of labor unions, and a much more restrictive land tenure system. But, for the most part, a "Yank" experiences little in the way of culture shock Down Under.

Australia often has been referred to as the "lucky country." Lucky to have been established by Britain when Britain was a benevolently successful colonial power. Lucky because it has been spared much of the tension and trouble that beset other parts of the world. Lucky because it had resources in abundance to undergird its development. Lucky because it was remote from the seats of power and, therefore, not a region to be fought over. Lucky because its Aboriginal inhabitants were few, unsophisticated, powerless, and easily displaced. Lucky because its populace has been relatively homogeneous and cohesive. Lucky because it never experienced revolt or revolution or war on its home territory. Lucky because it has never had to contend with population pressures.

Although lacking in water and large areas of fertile soil, Australia has a resource base that is more than adequate for agriculture. It is among the world's leading producers and exporters of wool, meat, wheat, and sugar. And the country is a dream for the economic geologist or mining engineer, because so many useful mineral resources have been found, and continue to be found, and still the subsurface geology is imperfectly known. Add to this the proximity of mineral-hungry countries in Asia (particularly Japan), Australia's comparative isolation from politically volatile areas, its internal political stability, and its relatively low inflation rate, and it becomes a very attractive place for large-scale mineral prospecting and development. Thus Australia is also among the world's leading producers and exporters of iron ore, coal, zinc, salt, alumina, gold, and diamonds.

The enormous output of the rural industries (pastoralism, farming, and mining) has been responsible in large measure for the fact that Australia's international reputation is rural although the population is overwhelmingly urban. The prosperity of the rural industries ebbs and flows, largely due to fluctuations in three "uncontrollable" factors—climate (which really means drought), markets, and labor problems. With the partial exception of the irrigation farmer, the man-on-the-land is at the mercy of the weather, and most parts of the country experience drought conditions frequently and lengthily. The state of the sky often determines the state of the bank account for pastoralists and farmers.

The second imponderable is the level of international prices and agreements, as the major products of farm, station, and mine are largely sold overseas, and the prevailing price is beyond the control of the Australian producer.

The third uncontrollable factor is labor. Although union membership has been slowly declining, Australia continues to be the world's most highly unionized country (40% of the labor force versus 20% in the United States), so that jurisdictional demarcation disputes resulting in work stoppages are commonplace. No other country comes close to matching Australia's record of work stoppages per capita. This factor affects all components of the economy, but is particularly notable in the mining industry.

Until the last quarter century, Australia was one of the great success stories in the economic history of the world. Indeed, the country has been called the finest triumph of long-distance colonization the world has known. Today, Australia continues to be one of the world's wealthiest countries and has one of the world's largest economies. It ranks in the top 25 nations in per capita income and Gross Domestic Product. Yet its high standing on the world economic ladder has been slipping in recent years, mostly due to low rates of GDP growth.

There is a continuing, and increasing, conflict between the traditional Aussie irreverence for the land and a growing concern for the integrity of the environment/ecology. The combination of vast area, dry climate, and sparse population engendered a frontier mentality that is still prevalent. This proclivity, however, is increasingly counterpointed by the arousal of a conservation ethic that burgeoned in the 1980s. "Development" is often important—sometimes essential—and development is still the main train of thought in Australia, but more and more attention is now being paid to the broader aspects of an undertaking, and it is abundantly clear that most Australians no longer take their environmental birthright for granted. Indeed, a highly publicized accord has been reached between various producer organizations and the Australian Conservation Foundation that augers well for the future. In connection with this agreement the 1990s has been declared as the "decade of Landcare" in Australia.

Perhaps the longest-running debate in Australian history has to do with immigration: Does the country need more people, and if so, how many and what kind? In comparison with the other continents, Australia has always seemed underpopulated, and the Australian government has devised many schemes to foster and subsidize immigration. Generally, it has been agreed that accelerated immigration is a good thing because it stimulates the economy by expanding both the labor force and the market as well as diversifying society, although in the last few years voices have been raised to question that assertion, noting that the ultimate fate of Australia is to become overpopulated through normal events, so why rush it? How many people can a desert continent support before the quality of life diminishes markedly?

The debate usually centers on the quantity of immigrants and their characteristics. Generally speaking, the numbers fluctuate in direct response to the governmental immigration policy of the moment, which is based on a complex suite of factors. In recent years, the annual net immigration has varied from a low of about 45,000 to a high of approximately 120,000.

Great Britain has always been the principal source of migrants, although in recent years the British proportion has diminished considerably. In the earlier days, the infamous "White Australia" policy inhibited (but did not prohibit) immigration from Asia, but that situation has long since been changed. In the 1980s, twice as many settlers came to Australia from Asia as from all Europe, including the British Isles. This rapidly increasing Asian component of the migrant total is a major focus of the current immigration debate.

As the ethnic mix of the Australian population broadens, the homogeneity of society diminishes. The result is either healthy multiculturalism or divisive ethnicity, depending on your point of view. In comparison with most other countries, the "plight" of minorities in Australia is not very severe, although the rapid recent influx of refugees and other migrants from Southeast Asia has resulted in some difficult situations.

By far the most prominent minority "problem" in Australia concerns the Aborigines. They are generally at the bottom of the socioeconomic scale, are plagued by unemployment and alcoholism, and have the highest school dropout rate. Their presence is very obscure to other Australians, however, because their numbers are small and the vast majority live in locations remote from the cities. In recent years, the Aborigines have developed a certain amount of political muscle, which was previously unknown, and have instituted claims of land ownership of large rural tracts, many of which have been approved by the government of South Australia and the federal government (for the Northern Territory), but which are stridently opposed by the governments of Western Australia and Queensland. The 1992 Mabo decision has given a new impetus to the land rights campaign, which unifies and strengthens Aboriginal efforts as well as emphasizing an abrasive political and economic issue.

In international relations, Australia's world view has been necessarily fashioned by geographical isolation. Located half a world away from the mother country and alienated by culture and custom from nearby Asian neighbors, Australians have generally and understandably been much more concerned with domestic affairs than they have with international events. The principal exception for many decades was the country's close liaison with Britain, which was the colonizing homeland, the source of most imports, the purchaser of most exports, and the paternalistic guide in foreign affairs. However, as a result of World War II and Britain's eventual joining of the European Economic Community, this relationship changed. Great Britain disengaged from its distant obligations, and Australia was forced to reassess its historic diplomatic and military dependence; the importance of alliance with the United States became undeniable, although not as dogmatically as had previously been the case with Britain.

Alienation from Britain continues to grow. In the early 1990s, a major political effort to change Australia from a constitutional monarchy to a republic by cutting all formal links with British royalty developed. This movement has received considerable support from Australia's present Labor government, and is a matter of polarized opinion all across the nation. Although the outcome is not clear at this time, it seems likely that all formal political ties between the two countries will be dissolved within a few years.

Australia has fostered an ever-increasing rapprochement with various Asian nations recently, a pragmatic recognition of certain mutuality of interests with countries that are geographical neighbors. In essence, Australia is striving to assume a comfortable stance as a "semi-Asian" country, as demonstrated by actual and potential trade relationships and a broadened immigration policy.

Australia has always maintained a considerable interest in the islands to its north, northeast, and east. New Zealand, with its common heritage and history, has been a major focus of this interest. (It is still true that in most years New Zealand is one of the principal sources of migrants to Australia, and vice versa.) Papua New Guinea, the country nearest Australia, has a particularly familial relationship because until 1975 it was an Australian colony. Australian relations with other South Pacific island countries are generally less intimate, but, in some cases, there is significant involvement.

What can be predicted about Australia's future? By any objective standard, it is still the "lucky" country. It still has remarkable actual and potential resources; it has been spared the actuality of overpopulation (although there is increasing sentiment that the specter of becoming an overpopulated country is just around the corner); it has been able to maintain a generally high standard of living by finding markets for its products; it has good quality medical services and a high quality of life.

But such good fortune cannot be taken for granted. It is all too easy for a boom to turn into a boomerang. The present crisis in the wool and sugar industries is a case in point. At this writing, Australia suffers from its highest unemployment rate since the depression of the 1930s. Economic and social problems are becoming more common. It is not enough to be a relatively underpopulated, relatively unpolluted storehouse of riches for the world. Australia's transition into a postindustrial society and an intra-global economy is a difficult one. In comparison with most of the rest of the world, the Aussies have a good life—but not as good as it used to be.

CHAPTER EIGHT

New Zealand: The Land and the People

Southeast of Australia, across the Tasman Sea, is the island nation of New Zealand. This southernmost nation in the Eastern Hemisphere consists of two principal islands (named the South and the North), which are the 12th and 14th largest in the world, as well as the much smaller Stewart Island in the far south. It has an essentially mid-latitude location, extending for 1,000 north-south miles (1,600 km) between 34° and 47° south latitude. Its area (104,000 square miles or 270,000 km^2) is only one-thirtieth that of Australia, but its population (3.5 million in the early 1990s) is nearly six times as dense as that of its larger neighbor.

Although the heritage and background of most of its populace are much the same as the people of Australia, its environment is remarkably different and its economy is more circumscribed. New Zealand is a land of diverse and gracious beauty, with scenic splendors of many kinds. The typical landscape is a rugged mountain background to verdant green pastures that are covered with sheep. The people of New Zealand have a relatively high standard of living and a democratic government that leans toward welfare state concepts, but a long-continued recession has persuaded the government to begin to dismantle the welfare system. The natural resource supply is limited. There is a narrow base for industrialization, therefore, and the economy is dangerously dependent upon export earnings from livestock products.

Environmental Diversity

The broad homogeneity that characterizes the Australian environment is lacking in New Zealand. Whereas large parts of Australia have an aspect of sameness, with almost monotonous expanses of flat land and sweeping vistas, the New Zealand landscape typically is characterized by sloping land and dense vegetation. Thus macro-generalizations about the physical geography of New Zealand are only broadly pertinent, and a real understanding of the environment requires analysis of the micro-aspects of exposure and altitude.

Terrain: Dominance of Slope. Unlike the ancient and inert land mass of Australia, New Zealand occupies an unstable section of the earth's crust, a position along the "Pacific Rim of Fire" where mountain building and other forms of crustal disturbance have been relatively active during recent geologic eras. New Zealand sits athwart the collision zone where the westward-moving Pacific plate is being subducted beneath the northward-drifting Indian-Australian plate. The warped edge of the latter was thrust upward with massive faulting and volcanic activity, raising thick layers of marine sediments to great heights. Pleistocene glaciation and abundant rainfall have made more recent erosional and depositional contributions to the shaping of the topography. Not all of New Zealand is high and steep, but slopeland dominates the terrain, and the horizon almost everywhere is notable for its high degree of local relief.

Two conspicuous undersea mountain arcs come together in the North Island, one trending southeasterly from New Guinea and New Caledonia via the Northland Peninsula and the other southwesterly from Tonga and the Kermadec Islands into East Cape (see Figure 8-1). The mountain trends continue southwesterly the full length of the country, but the pattern of ranges is irregular and discontinuous for the most part and is repeatedly interrupted by many narrow valleys and a few broad plains.

The North Island consists of hilly country underlain by relatively weak rock, although there is a hardrock mountain core along the eastern side of the island. The surface expression of the core is mostly hill land consisting of several disconnected ranges, separated by small plains and basins. In the center of the island, there is a conspicuous volcanic region containing dissected plateaus, notable volcanic peaks, a complex of hydrothermal features, and the country's largest lake. Lake Taupo is nearly twice as large as any other lake in New Zealand, and its river system is easily the longest in the country, as well as being the principal source of hydroelectricity for the North Island.

On the south side of the lake are three prominent volcanic peaks, all active, whose summits range from 6,500 to 9,200 feet (1,950 to 2,760 m) above sea level.[1] Vulcanism on the north side of the lake is manifested by a multitude of geysers, hot springs, fumaroles, and geothermal steam vents. The northern portion of the North Island is characterized by more gentle terrain, with rolling country alternating with flattish plains in the Waikato District and the Northland peninsula. The dominant topographic feature of the western part of the island is the symmetrical cone of Mt. Egmont (whose

[1]Mt. Ruapehu, the tallest at 9,175 feet (2,750 m), has six small glaciers on its flanks.

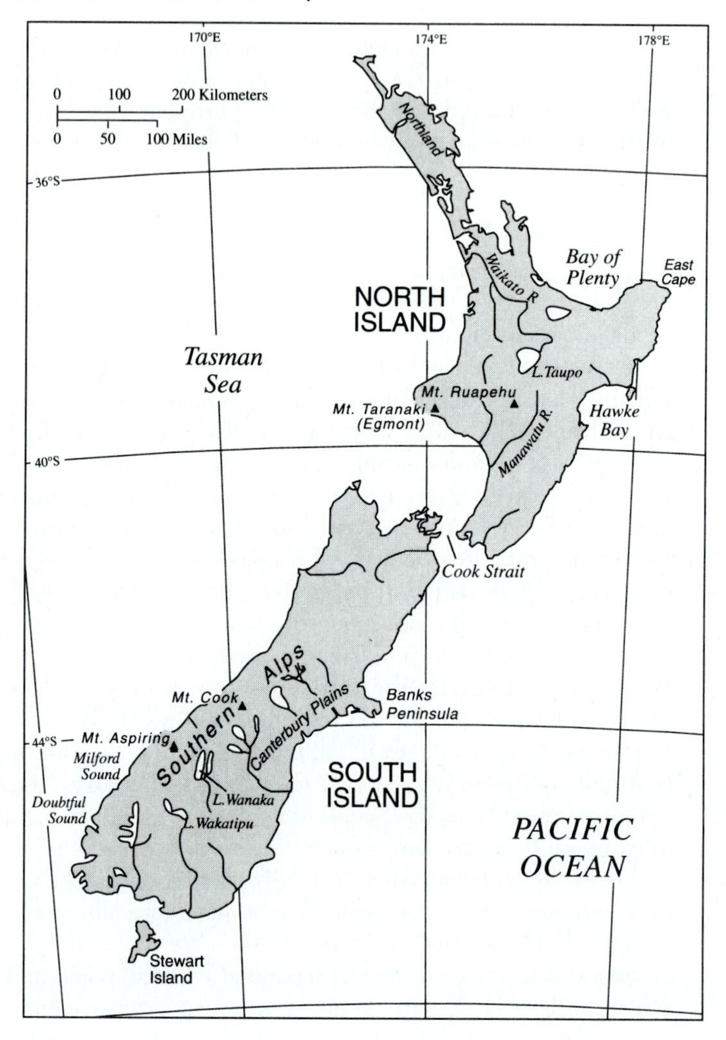

Figure 8-1 New Zealand's Prominent Physical Features

other official name is Mt. Taranaki), which rises abruptly to 8,300 feet (2490 m) above the fringing Taranaki lowland (see Photo 8-1).

The South Island is not only larger, but also more mountainous and rugged than the North Island. However, its volcanic manifestations are fewer, being restricted to ancient formations in the Banks Peninsula near Christchurch. The massive mountain chain known as the Southern Alps makes up the backbone of the island and occupies about half of its surface area. The highest and most spectacular portion of the Alps is in the center of the chain, particularly around Mt. Cook (12,350 feet or 3,700 m) and Mt. Aspiring (9,950 feet or 2,985 m), where extensive snowfields occur and lengthy glaciers reach down west side valleys almost to tidewater. There are

Photo 8-1 *Mt. Egmont's impressive cone towers above the Taranaki dairy country near New Plymouth. Remnant snowfields can be seen in this mid-summer (February) photograph. (TLM photo.)*

3,155 active glaciers in the Southern Alps, most of them within 100 miles (160 km) of Mt. Cook.

In the north of the island, the main chain of the Alps divides into a confused sequence of discrete ranges, most of which trend northeast-southwest and are separated by longitudinal valleys. These ranges are lower, unglaciated, and less rugged. The southern portion of the Alps also breaks up into a mixed pattern of ranges and valleys; elevations are lower there, but the mountains have been carved into a remarkable series of steep-sided valleys and fiords by past glacial action. The western slope of the entire chain of the Southern Alps is marked by an abrupt, often precipitous, descent to the coast, with almost no coastal plain.

On the eastern side, in contrast, there are extensive foothills, longitudinal glaciated valleys (many of them occupied by deep freshwater lakes), vast areas covered with *shingle* (glacio-fluvial outwash of sand and gravel), and a number of intricately braided rivers. In the east-central portion of the South Island is the Canterbury Plains, the only extensive lowland area in all New Zealand. It is a piedmont alluvial plain veneered deeply with shingle and crossed by many southeast-flowing rivers. The southeastern portion of the South Island is an irregularly dissected section comprising the Otago and Southland districts. Otago is mostly hill country, and Southland contains a number of small plains and basins separated by hills.

Climate: A Well-Watered Land. New Zealand's climate is the type classified as Marine West Coast, that is, a mid-latitude climate dominated by moist air from the ocean. Only on the east (leeward) side of major mountains is there sufficient protection from oceanic air to produce a relatively dry environment. The principal climatic determinants are latitude, altitude, exposure, prevailing winds, and the maritime surroundings. The mid-latitude location of New Zealand dictates that westerly winds generally sweep the islands, bringing a year-round pattern of irregular cyclonic storms and meridional fronts.

Precipitation is moderate to abundant over most of New Zealand; average annual figures of 30 to 60 inches (750 to 1,500 mm) typify most of the settled parts of the country. Mountainous areas receive more moisture; much of the high country gets more than 100 inches (2,500 mm) annually, and a fairly extensive section of the west side of the Southern Alps records in excess of 200 inches (5,000 mm). The "rain shadow" position of the Canterbury and Otago districts, on the leeward side of the Alps, inhibits precipitation so that many places receive less than 25 inches (650 mm) annually and a few localities get less than 15 inches (380 mm). The seasonal regime of precipitation is fairly regular throughout the year, with a tendency toward a winter maximum on the North Island and a slight summer maximum on the South Island. Snowfall is uncommon in the lowlands, but is quite heavy in the mountains of both islands.

Temperatures are generally moderate except in the mountains. In most settled parts of the country, one may expect summer temperatures to range between 50° and 80°F (10° and 27°C), with winter temperatures varying between 30° and 70°F (-1° and 21°C). No significant town has ever experienced a temperature above 100°F (36°C) or below 10°F (-12°C), although much lower minima can be expected in the mountains. The normally high level of relative humidity tends to emphasize the cooler temperatures rather than the warmer ones.

Windiness is a widespread characteristic of New Zealand weather. Both Wellington and Dunedin are notoriously windy cities, and the waters of Cook Strait that separate the two main islands are almost continually windblown. The passage of fronts and other types of pressure systems bring frequent blustery conditions, particularly to the South Island.

Although temporary exceptions are numerous, New Zealand's typical atmospheric conditions are moderate. Both annual and daily temperature ranges are muted by the oceanic influence, except at higher elevations. Precipitation is relatively frequent, but storms normally do not linger.

Natural Vegetation: Variety in Abundance. The natural vegetation patterns of New Zealand are more complex than those of Australia, with much greater local variations in distribution. Furthermore, the plant communities of New Zealand have been relatively unstable, that is, in a state of change, during recent geologic eras because of vagaries of nature (volcanic activity and glaciation) and within historic time because of human activities (fires, heavy grazing, and land clearance).

Prior to human interference, the islands were heavily forested; more than three-quarters of the total area was covered with what the early European settlers called *bush*, much of which was actually a sort of temperate rainforest. Massive kauri (*Agathis australis*) trees dominated in the milder north and extensive stands of southern beech (*Nothofagus* sp.) typified the cooler south. The remaining forests are classified broadly into either mixed temperate evergreen forest or *Nothofagus* forest.[2] The former is a varied community of many species of broadleaved trees and conifers, and the latter a relatively pure community of one or more of the species of southern beech. Generally, the mixed forests are typical of the north and of the wet lowlands and lower mountain slopes whereas the beeches are mostly in the south, on the high mountains, and in the drier lowlands. However, the two types frequently intermingle.

Throughout the country there were many areas dominated by scrub and ferns. The boundaries between forest and scrub were both irregular and subject to rapid fluctuation; with a respite from burning the forest advanced, and with increased burning the scrub advanced.

Some large areas were grass covered, although much of this apparently was a result of frequent burning by the early Polynesian inhabitants. Tussock grassland was common east of the Alps in the South Island and in some of the volcanic heartland of the North Island. Alpine grasslands, or tundra, were characteristic above the treeline on higher mountains.

Much of the native vegetation has been removed or replaced in recent years; partially by clearing for agriculture, but much more significantly by firing and overgrazing. Most of the pasture land of New Zealand today has been seeded to exotic grasses introduced from various parts of the world.

Fauna: Scarce or Exotic. Whereas the native animal life of Australia can be characterized as unique and bizarre, that of New Zealand is best described as "absent." Indigenous terrestrial animals were almost entirely missing; there were no land mammals at all, no reptiles except for a few species of lizards, no amphibians except for a few species of frogs.

Avian fauna, on the other hand, was conspicuous and dominant. Waterfowl were the most common, particularly in coastal areas, but there were many other varieties. A distinctive feature of the bird life was the relatively high degree of flightlessness. The kiwi (New Zealand's national emblem) shares this characteristic with a variety of rails, woodhens, and other species, but the most notable of non-flying birds was the moa, which finally died out

[2]In fact, both of these categories of flora are "evergreen." New Zealand has no native deciduous trees.

just before the European "discovery" of New Zealand. The largest of the two dozen species of moa grew to a height of twelve feet (3.6 m).

The absence of native animal life encouraged the European settlers to import exotic forms from many parts of the world. Indeed, New Zealand has become the world's most notable example of a land wherein exotic animals dominate the fauna. Some 30 species of birds, 15 species of mammals (most notably rabbits and various ungulates from Europe, and marsupials from Australia), and a number of fish from other continents have become established members of the New Zealand biota, generally to the detriment of both the ecology and the economy of the country. Feral livestock, particularly pigs and goats, add to the faunal confusion.

The Human Occupance

The prehistory of New Zealand is still imperfectly understood. From the limited archaeological evidence that is available, we can infer that the magnificent landscape of these islands was untrammeled by human footsteps until much later than almost any other portion of the earth's surface. It is believed that New Zealand remained unoccupied by humans until well after the beginning of the Christian era.

Pre-European Settlers. The first humans to settle in New Zealand apparently were migratory Polynesians who arrived between 1,000 and 2,000 years ago. The precise date is unknown, but there is no certain evidence of settlement before about AD 800. Their origin was somewhere in eastern Polynesia, although the specific islands from which they came is also unclear. Apparently, the migration was purposeful, as it included both men and women, and their cultural baggage included several crop plants, various tools and implements, and dogs. The scanty available evidence indicates that they tended to concentrate in east coastal settlements on the South Island. They were peaceful, unsophisticated people whose economy was based essentially on bird hunting and fishing. A principal quarry was the giant moa, and the people have often been referred to simply as "Moa-hunters," although the Maoris called them Morioris (primitive people), and some scholars prefer the term "Archaic Maoris."

The relationship between the Archaic Maoris of earlier times and the Classic Maoris who occupied New Zealand at the time of European contact is also unclear. It is probable that the Moa-hunter culture developed into the Classic Maori culture through a normal process of cultural evolution. From time to time, the population mix was enriched by the abrupt addition of newcomers, usually Polynesian immigrants arriving from islands far to the north or northeast. There is considerable evidence that a final wave of immigration (the Great Fleet) arrived during a relatively short time span in the 14th century, giving notable external stimulus to the development of the Classic Maori culture.

In any event, the Morioris had been assimilated or eliminated by the time of European contact (late in the 18th century), and the Classic Maori society was well established. At that time, their numbers were estimated to be between 100,000 and 200,000. More than 80% lived in the northern half of the North Island. Their economy was significantly agricultural, based largely on shifting cultivation, with the kumara (sweet potato) as the principal crop. They transformed the kumara from a tropical perennial into a temperate seasonal crop that could be grown as far south as the middle of the South Island. Storage in underground pits enabled the harvest to be saved for future use. A few other crops—such as taro and yams—were also grown, but they were not nearly as important as the kumara. The Maoris also spent much time bird hunting, fishing, and gathering such edible native plants as fernroot. Their art forms were limited; they were skilled woodcarvers, and were noted for their proficiency in tatooing, flax-working, and military engineering.

The Classic Maoris lived in scattered hamlets and in fortified villages (called *pa*). Remains of some 5,000 *pa* have been identified on the North Island, but only about 100 on the South. They were divided into a large number of politically autonomous tribes, each with a more or less recognized territory. The tribes generally were grouped into a number of loosely knit associations, and intergroup warfare was frequent, if not incessant.[3]

Most of the remainder of the Maori population settled in the southern half of the North Island and along the north coast of the South Island. They were more dependent upon hunting, fishing, and gathering than they were upon agriculture; their settlements were generally smaller and less permanent; and they were less warlike.

Most of the South Island was very sparsely settled, although some Maori settlements were found as far south as Stewart Island. The South Island Maoris had a primitive subsistence economy.

Arrival of the Europeans. First on the scene were Abel Tasman's two Dutch ships in 1642, which were on a vast circumnavigation of Australia from Java. Tasman observed little and recorded even less of the islands he called "Staaten Land" (but which the Dutch subsequently christened "Nieuw Zeeland"). Indeed, he attempted only one landing, which resulted in the fatal spearing of several of his men by Maoris. No other Europeans visited New Zealand for 127 years. In the late 1700s, several British, French, and Spanish vessels came.

James Cook was the first European explorer to land in New Zealand. On his first Pacific voyage, Cook spent six months in 1769 and 1770 circumnavigating both main islands and carefully mapping their coasts. Cook had generally peaceful dealings with the truculent Maoris, and returned to New

[3]Joseph Banks, botanist with the Cook expedition, commented on the warlike proclivities of the Maoris: "I suppose they live entirely on fish, dogs, and enemies." (Robert Hughes, *The Fatal Shore* [New York: Alfred A. Knopf, 1986], p. 52.)

Zealand on three other occasions during the 1770s. As the Maoris kept no written records, the recorded history of New Zealand dates back only to Cook's time.

The first European settlement was a sealing station established on the west coast of the South Island in 1792, and several other sealing and whaling stations were opened during the subsequent four decades. These temporary settlements became more permanent and more numerous, although New Zealand still was not officially claimed by any colonial power. Deserters from ships and escaped convicts from Australia made up a significant proportion of the small European population; lawlessness was rampant, and there was sporadic fighting with the Maoris. Some significant trading of whale oil, seal skins, timber, and flax was carried on, particularly with New South Wales.

Missionaries and others were eager to establish viable settlements, but all of the early attempts failed. Finally, in 1840, the first true colonization settlement was founded at Wellington by the New Zealand Company, a British enterprise whose guiding hand was Edward Gibbon Wakefield, who had also led the South Australian Company that had founded Adelaide. In that very same week, British sovereignty was proclaimed over New Zealand. Captain William Hobson of the British Navy (soon to become the first Governor of New Zealand) and a group of Maori chiefs signed the Treaty of Waitangi, which acknowledged British rule over the North Island. The South Island and Stewart Island were proclaimed to be under British sovereignty by right of discovery. The Treaty of Waitangi provided that the Maoris accept Queen Victoria as their ruler in return for British protection of all Maori rights, including property rights.[4] New Zealand was governed at first as a dependency of New South Wales, but, after a year, it was proclaimed a separate colony by Royal Charter.

By 1850, there were six separate settlements in New Zealand—four on the North Island and two on the South. By that time most of the Maori tribes had managed to obtain guns, so their intertribal wars became more deadly. It is estimated that they managed to reduce their population, largely through internecine warfare, by 40 or 50% during the second quarter of the 19th century.

The Spread of Settlement. The second half of the 19th century in New Zealand, as in Australia, was a time of rapid population increase and settlement expansion. Furthermore, during this period the tide of settlement shifted from the North Island to the South Island and back again.

In the early 1850s, the North Island population, both *pakeha* (the Maori term for Europeans) and Maori, was considerably greater than that of the South Island. The Europeans, with only a handful of exceptions, were concentrated on the coasts. The interior was largely Maori country; both

[4]There are two versions of the treaty; one in English and the other in Maori. Problems with the translation and implication of such important words as "sovereignty" and "governorship" have led to serious disputes which continue today.

danger from the Maori and lack of economic stimulus served to discourage European penetration inland. The introduction of sheep was the first factor influencing a change in the pattern. The prospect of wool production lured some hardy settlers away from the coast, but a relatively few men claimed most of the land.

Gold was discovered in inland Otago in 1861, signaling a major reorientation in both the economy and demography of New Zealand. There was a great migration to the South Island from the North Island, from Australia, and from overseas. Within less than a decade Otago had one-fourth of New Zealand's population, and its principal town, Dunedin, numbered some 60,000 inhabitants, making it by far the largest settlement in the colony.

Meanwhile, on the North Island there were increasing attempts at inland settlement by Europeans. The alienation of Maori land brought about a sharp increase in hostilities with the pakeha, and the entire decade of the 1860s was marked by the "New Zealand Wars" (also called the "Maori Wars" or "Land Wars"), which resulted in inevitable defeat of the Maoris (except for some who sided with the government) and a rapid diffusion of European land ownership in the interior of the island.

European population growth and settlement expansion were more rapid in the South Island, however, as there was only a small Maori population to deal with. The Canterbury Plains became an important agricultural area in the 1870s and 1880s, emphasizing wheat.

Perhaps the most important event in the entire economic history of New Zealand was the inauguration of refrigerated shipping in 1882. Until that time the principal exports had been such nonperishable items as wool, tallow, wheat, and gold. With refrigeration, however, meat and dairy products could be added to the list. Whereas most New Zealand sheep had been Merinos, raised strictly for their wool, mutton and lamb could now be marketed. Accordingly, dual-purpose breeds became more important, with the varieties known as Romney Marsh (imported from Britain) and Corriedale (developed in New Zealand) soon dominating.

Much of the interior of both islands, apart from that continually diminishing area still under Maori control in the North Island, was held in a relatively few large estates. This pattern of land tenure came under increasingly severe pressure. Squatters settled in many places with little consideration of the legality of their claims. However, some of the large estates were broken up voluntarily, and legislation in the 1890s provided for compulsory subdivision of most of the remainder. The relatively dense growth of forest and bush over much of the North Island was an impediment, but vast acreages were cleared in a relatively short time. Land clearance was less significant on the South Island.

In terms of population concentration, the pendulum swung north again, probably never to change. The rate of population growth on the North Island was faster than that on the South during the last two decades of the 19th century, and by the dawn of the 20th century the North Island contained more than half of New Zealand's total population. The spread and intensification of settlement continued to be faster on the North Island, depending primarily upon fat-lamb raising and dairy production of butter and cheese for export. Early in this century, most of New Zealand's usable land was occupied.

The Populace

In the early 1990s, the population of New Zealand was about 3,400,000. It is growing at a rate of slightly less than 1% per year, an annual gain of some 30,000 people. Natural increase has accounted for all of the net population growth since about 1975, because there has been an annual net outflow of emigrants of between 10,000 and 40,000 per year, reflecting New Zealand's economic distress. The demographic characteristics of the population (birth rate, death rate, age structure, and sex ratio) are very similar to those of Australia, with some significant differences in ethnic composition.

Following a trend that began about a century ago, the population continues to concentrate on the North Island. Today, more than 75% of all New Zealanders live there, in comparison with less than 50% at the turn of the century. The northern half of the North Island is experiencing a rate of increase twice as fast as the rate in the southern half of the North Island. Indeed, more than one-fourth of the total population now resides in metropolitan Auckland.

Most of the population lives in cities and towns, in a ratio roughly the same as for Australia. Nearly 85% of the population is classed as urban; the urban proportion is higher on the North Island than on the South. There is distinct rural depopulation in some areas, with many rural counties and towns on a downward trend; a situation unlike that in Australia.

Five major urban areas contain nearly half of the total population, and are significantly larger than any other cities. Metropolitan *Auckland* has a population of more than 850,000. It is the principal commercial and industrial center of the nation, has the busiest port by far, and is the only New Zealand city with the cosmopolitan characteristics of similar sized cities in other countries. *Wellington* is the national capital and, with its suburbs in the Hutt Valley, sprawls around three sides of the spacious bay called Port Nicholson. The combined population of Wellington/Hutt is approaching 400,000. *Christchurch*, the principal city of the South Island and hub of the fertile Canterbury Plains, contains a population of about 300,000. *Dunedin*, the long established commercial center of the Otago District, has about 110,000 residents. *Hamilton*, in the fertile Waikato Valley, is the fast-growing newcomer to the list of major cities, with a population approximately equaling that of Dunedin. Three dozen other places are officially classed as "urban areas" in New Zealand, but only one of them is even half as large as Dunedin or Hamilton.

Until recently, there has been a strong flow of immigration to New Zealand since World War II, with the exception of certain bad economic years. As in Australia, the national government sets annual immigration goals and subsidizes many of the migrants. The United Kingdom has been by far the major source—more than 60% of all migrants to New Zealand since World War II have been from Britain. The second largest inflow has been from Australia, but the outflow to Australia roughly balances that in most years. The second largest *net* source of migrants has been the islands of Polynesia, especially the Cook Islands, Western Samoa, Tonga, Niue, and Tokelau, all of which (except Tonga) are governed or were previously governed by New Zealand.

The Maoris comprise a small but significantly minority (about 12% of the total population), and are easily the most conspicuous non-European element. At the beginning of the century, there were only about 40,000 Maoris; today the total is more than 10 times as great. Their high birth rate yields a net rate of increase that is twice that of the general population. More than 90% of the Maoris live on the North Island, and their rate of urbanization is higher than that of the total population.

The Maoris have equal rights with other New Zealanders, but have a generally low economic position in society. The historic dispute between Maoris and *pakehas* over property rights, dating from the Treaty of Waitangi, has become a prominent political issue today. The government has established a special judicial body, called the Waitangi Tribunal, to consider Maori claims.[5]

A Narrowly Based Economy

New Zealand's economy is dominated by its rural industries, especially animal husbandry. Although the majority of the labor force is employed in secondary and tertiary activities, these are supported by a narrow resource base. Sheep raising, dairying, and beef production provide the undergirding of the economy. The domestic market is not large enough to absorb all of the butter, wool, mutton, lamb, beef, veal, and cheese produced; thus, overseas sales take about two-thirds of the output.

Gross farm income is divided approximately as follows: one-half from pastoral products, one-third from dairy products, and one-sixth from crops. Moreover, more than 50% of the nation's exports consist of animal products. In a real sense, then, New Zealand's economy is dependent upon the nutritious carpet of pasture grasses that have been introduced to replace the indigenous vegetation. The country's economic history has been dominated by the prosaic but continuing struggle to clear the bush and plant new grasses.

[5]The Tribunal, created in 1975, investigates claims made by Maoris who feel they have been unfairly dealt with by the government in an action or omission inconsistent with the *principles* (not just the text) of the Treaty of Waitangi. The Tribunal has the sole right to determine the meaning and effect of the treaty where differences in the Maori and English texts occur, but the Tribunal's judgments are non-binding.

The leading rural industry is sheep raising (no other country can claim 20 times as many sheep as people). New Zealand's 70,000,000 sheep are widely spread, but the greatest concentrations and the more intensive activities are on the North Island, where some 55% of the total is found. Most North Island sheep farms are relatively small (1,000 to 3,000 acres/400 to 1200 ha), graze a few hundred or a few thousand head of sheep on improved pastures, and produce both wool and meat as major products. In many cases, particularly in the northern part of the island, sheep raising and dairying are carried out on the same farm. On the South Island, nearly half of the sheep are maintained on extensive, high-country "runs," where woolly Merinos are dominant (see Photo 8-2). On the Canterbury Plains and in some of the smaller lowlands of the South Island, large numbers of sheep are kept in mixed farming enterprises, which are similar in many ways to operations in the Australian wheat-sheep belt.

When Great Britain joined the European Economic Community, it spelled hardship for the entire New Zealand economy because the favored status of British imports of New Zealand products was phased out. The rural industries were hardest hit. As sheepmeat sales to the United Kingdom declined, much diligence was exerted in seeking alternate markets, with

Photo 8-2 *Merino sheep being overlanded in the Crown Ranges on the South Island. (TLM photo.)*

some compensating results in the Middle East, Japan, and Latin America. Wool sales were never so concentrated in the United Kingdom, and there have been erratic marketing gains in the Far East and in Russia.

The 8,000,000 cattle in New Zealand are divided approximately 60%:40% between beef and dairy types. Dairying, however, is second only to sheep raising among the nation's leading agricultural activities. It is carried on over most of the country except in the higher mountains, but its major stronghold is in the northern third of the North Island, from the Bay of Plenty coastland and the Waikato District northward. Most of the dairying country was originally forested; the productive pastures of the present are almost completely the result of clearing the land and planting exotic grasses.

In recent years, the New Zealand dairy industry has been broadened from its traditional triumvirate of milk-butter-cheese to include wholemilk powder, anhydrous milkfat, casein, whey, and other specialized products. Nevertheless, the viability of the dairy industry continues to be based upon the long-established export of butter and cheese to Britain. The British market has been progressively reduced, but it still absorbs about half of the butter exported from New Zealand. That New Zealand has been able to maintain its position as the world's leading exporter of dairy products is due partly to assiduous cultivation of a variety of (often undependable) markets in other parts of the world.

New Zealand's 5,000,000 beef cattle are widespread over the North Island, which pastures four-fifths of the total, but occur in much more limited locations and density in the South Island. Most beef cattle producers also raise other types of livestock (sheep, dairy cattle, or deer). In recent years, there has been a downward trend in the beef cattle inventory, due to declining overseas markets.

Other agricultural activities of significance are the growing of crops and the raising of pigs, deer, goats, racehorses, and chickens. Most crops are grown in mixed crop-and-livestock operations which have meat as a significant farm product, except in the Canterbury Plains where specialty grain farms are common. The South Island contains nearly three-fourths of the total crop acreage, particularly in the Canterbury Plains and the coastal lowlands of Otago. The major cash crops are wheat, oats, barley, and potatoes, and much acreage is devoted to the growing of such items as turnips, swedes, and peas for stock feed; however, the majority of the total farmland acreage is sown to grasses and clovers for grazing.

There is also a certain amount of market gardening and orcharding, particularly around Auckland and Hawke Bay on the North Island and around Nelson on the South Island; these enterprises produce a wide range of vegetables, fruits, and berries, largely for domestic consumption; the only New Zealand crops that are exported significantly are kiwifruit and apples, the former a recent development that already is feeling the pinch of competition from newer producers in various Asian countries. The vineyard/wine industry is also expanding rapidly, particularly in the Gisborne, Hawkes Bay, and Marlborough areas.

Focus Box: Deer Farming

One of the most intriguing recent developments in Antipodean agriculture is the rise of deer farming. Deer have been a part of the "wild" rural scene in New Zealand for many decades, as they were imported and liberated on numerous occasions in the 19th century by "acclimatisation societies" who felt that the lack of native mammals was a problem that should be rectified by direct action. About a dozen different varieties were introduced; mostly from Europe, but a couple of species each from Asia and North America. A few species died out, but most established viable wild populations and some became so numerous as to become pests.

Most successful by far in adapting to the New Zealand environment was the European red deer, *Cervus elaphus.* It prospered almost everywhere in the country and, in many places, was considered a scourge as obnoxious as the rabbit. New Zealanders were encouraged to kill deer, especially red deer, anywhere, anytime. Beginning in the early 1930s, the government employed several dozen "deer cullers" whose full-time jobs were simply to shoot deer. Eventually helicopters were used extensively by the cullers, who have made nearly one million deer kills in the last half century (a number that does not include the harvest of private hunters).

Venison began to be exported in the early 1950s, and soon a flourishing trade was established, particularly with West Germany. High prices for venison attracted the attention of farmers with deer running on their properties or in adjacent bush country. Before long, there was considerable interest in pasturing deer under controlled conditions. Deer farming was legalized in 1969.

In the 1980s, deer farming suddenly became big business in New Zealand. There are now more than 400,000 deer kept on some 3,500 farms (see Photo 8-3). They are grazed, much like sheep, in expansive but tightly fenced paddocks; indeed, many deer farmers raise sheep as well. Ninety percent of the New Zealand farmed deer are red deer; most of the others are either American elk (wapiti) or European fallow deer. Tax regulations encourage expansion, and it is not uncommon for a vigorous male to be sold for $15,000 or $20,000.

The principal product of the industry is meat—venison has become a rapidly growing specialty export from New Zealand, especially to Australia. The other commodity of significance is soft new antler (called "velvet" in the trade), which is surgically removed after it has been growing for two or three months and sold, mostly in Asia, for medicinal (largely aphrodisiac) use. High quality velvet retails for about $50 a pound (.45 kg). At present, deer farming is one of the brightest spots in the New Zealand agricultural scene.

The story of deer in New Zealand is a very revealing saga of the unexpected permutations that can result when humans manipulate the environment in helter-skelter fashion. First a void, then an interesting new faunal component, then a nuisance, then a pest, and now a valuable agricultural commodity—all within a century. What will be the role of New Zealand deer a few years hence?

Photo 8-3 *A red deer farm on the slopes overlooking Lake Wakatipu near Queenstown on the South Island. (TLM photo.)*

New Zealand is not a wealthy country in terms of natural resources, but electric power generation is one of the strengths in its resource inventory. The high elevations, glaciated history, steep slopes, and abundant rainfall produce many natural lakes and swift streams, which offer a high potential for hydroelectric power. The development and utilization of this potential (about 75% of the nation's electricity comes from this source) has been imaginatively and extensively accomplished, so that a first-class, interconnected power grid blankets both islands, and the per capita consumption of electric energy in New Zealand is one of the highest in the world. The Waikato River is the historic producer of most of the nation's hydroelectricity, and it is still the leading producer on the North Island. In recent years, however, large facilities have been constructed on the South Island (especially on the Rakaia, Waitaki, and Clutha rivers and on Lake Manapouri in Fiordland), which far eclipse output from the North Island.

Electricity also is generated from New Zealand's relatively abundant but low-grade coal supplies, which are mostly in the Waikato basin of the North Island. Natural gas is used as a minor but increasing source of electricity production, and one of the world's largest projects for the production of electricity from geothermal steam operates in the Wairakei area just north of Lake Taupo.

The complex geology of this island nation provides a wide variety of rock types, but only a few economic minerals are available in worthwhile amounts. Coal supplies are adequate to fuel a modest steel industry, and the blacksand beaches of the western coasts of both islands contain a sufficiency of iron particles to provide the needed ore and even to allow a small surplus for export. Crude oil has been produced in trivial quantities for more than a century, and production is slowly increasing. Natural gas, on the other hand, has been discovered in large amounts off the west coast of the North Island, and New Zealand is now self-sufficient in gas output.

Originally, New Zealand was heavily forested. It is estimated that more than three-fourths of its total area was tree-covered prior to the first human settlement, and more than 50% was still forested when the *pakeha* arrived. Widespread clearing, for timber sales and for expansion of pastoralism, reduced this by half by the early years of the 20th century. Since then, the remaining native forests have been managed much more conservatively, but the major forestry development has been in the establishment of more than two million acres (800,000 ha) of exotic softwood plantations, mostly on the central volcanic plateau of the North Island. The principal species, by far, is the Monterey pine, *Pinus radiata*, from California. These plantations yield more than nine-tenths of the nation's total cut of sawn timber, provide the basis for an expanding pulp and paper industry, and supply an increasing export trade to various Asian markets.

Commercial fishing in New Zealand has an erratic and generally low-profile history, despite a variety of marine resources in the surrounding seas. In the last decade or so, however, vigorous expansion, stimulated by government assistance for modernization and rationalization, has taken place, and the industry continues to grow. By the early 1990s, fishing contributed as much as forestry to the country's export revenue.

Secondary and tertiary activities are very important, but there is little distinctive about their development. New Zealand possesses the range of commercial and service industries that should be expected of a small country with a high standard of living. The scope of these activities is strictly limited to satisfying the domestic demand.

Manufacturing industries are similarly limited. Although some 300,000 people are employed in manufacturing (amounting to more than one-fifth of the labor force), this segment of the economy has depended very heavily upon government encouragement and protection for its growth. Protective tariffs, customs duties, and import restrictions have made it possible for a wide range of manufacturing industries to develop, many of which would not be viable in an unprotected economy. Like Australia, however, since the mid-1980s many tariffs have been reduced and other types of deregulation have been introduced, so that New Zealand manufacturers now face increased foreign competition. Even so, the industrial structure shows the fairly restricted role of secondary industry. The principal type of manufacturing, by far, is the processing of agricultural produce, which employs almost 25% of the entire manufacturing work force. In recent years, there has been considerable growth in the machinery and metal fabricating industries. Other leading types of manufacturing are wood products, transportation equipment, pulp and paper products, and printing and publishing.

In terms of location, there is a close correspondence between the distribution of factories and the distribution of urban population. Metropolitan Auckland contains more than one-third of the nation's factories; Wellington and Christchurch have about one-sixth each. The North Island continues to experience considerably more industrial development than does the South. Overall, the manufacturing segment of the economy is becoming increasingly important.

New Zealand's transportation infrastructure is well-developed and efficient, despite the engineering handicaps bestowed by the pervasiveness of hilly and mountainous terrain. The 60,000 miles (96,000 km) of roads have an average of one bridge every three miles, and the 2,700 miles (4,300 km) of railway require almost one bridge per mile as well as a total of 165 railway tunnels. A comprehensive network of roads and railways covers most of the North Island and the eastern portion of the South Island. Cook Strait, between the two islands, is crossed by several roll-on, roll-off ferries that accommodate both motor vehicles and railway cars. As in Australia, passenger movements are overwhelmingly dependent on automobiles and aircraft. Motor vehicle ownership (1.6 persons per vehicle) and per capita use of air transport are among the highest in the world.

One of the bright spots in the New Zealand economy in recent years has been the development of an international tourist industry. The image of an attractive country with magnificent scenery and friendly people is indeed a reality. The main drawback is the vast distance separating New Zealand from sources of tourists. However, visitation from Australia, the United States, Japan, Britain, and Canada has been expanding rapidly. Tourism is now New Zealand's largest single foreign exchange earner, exceeding the traditional primary products of the pastoral and dairy industries. The four focal points of New Zealand tourism are the city of Auckland, the geyser area of Rotorua, the high country around Mt. Cook, and the magnificent Fiordland scenery around Milford Sound. Some one million tourists per year were visiting New Zealand in the early 1990s, twice the number of a decade earlier, with Australia and the United States as the leading sources.

Throughout its short history, New Zealand has been heavily dependent upon foreign trade for its development; its per capita value of foreign trade is still one of the highest of any country. The trading patterns have always been distinctly "colonial," involving the export of primary foodstuffs and raw materials and the import of manufactured goods. Until the 1960s, Great Britain was both the principal customer (taking about 80% of exports) and the leading supplier (providing more than 50% of imports).

When Britain joined the European Economic Community, however, it was a requirement that New Zealand's position as a favored supplier of primary products be severely reduced. This was a staggering blow to the New Zealand economy, as Britain was the principal buyer of all of New

Zealand's major exports—butter, cheese, lamb, beef, veal, and wool. Enormous efforts have been expended to develop both replacement markets and diversified products, with modest success. Japan is now New Zealand's largest trading partner, and trade volume has increased significantly with the United States, as well as several countries of western Europe and the Middle East. Great Britain (at a greatly reduced level) and Australia continue to be important trading partners for New Zealand.

Problems and Prospects

The overriding consideration in any assessment of New Zealand's future is the perilous position of the economy. Heavy dependence upon export of primary products to balance the steady flow of needed imports and to energize the domestic economy has always been a delicate matter and has required astute governmental management. In recent years, this "management" has had two major challenges:

1. to develop replacement markets for the diminishing role of Great Britain; and
2. to diversify exports away from heavy dependence on products of the pastoral industry.

The retreat from the British umbrella was much easier than anticipated in the 1970s because the world economy was generally buoyant and other buyers were not so difficult to find. In the 1980s, however, recession conditions narrowed the opportunities and it seems clear that New Zealand faces difficult times in retaining its customers, much less in expanding its markets.

The government has done much to encourage diversification of production for export, with surprisingly encouraging results. Whereas 20 years ago, non-pastoral products comprised only 15% of total exports, today that share is closer to 50% with much of the increase coming in the form of wood products, minerals, and manufactured goods. Many of these gains, however, are tenuous, and depend upon government subsidy, which is being phased out.

There is significant disparity in regional development in New Zealand. Most of the economic activity is urban-oriented, and many rural areas are declining or stagnating in both population and income. Moreover, Auckland's dominance continues to grow, and with it comes an acceleration of the common metropolitan problems of the Western world—congestion, pollution, crime, stress and social dislocation.

Despite widespread economic problems, New Zealand has made one of the strongest environmental protection commitments of any country. Its Resources Management Act, promulgated in 1991, enshrined the concept of sustainable management into all of the nation's resource and planning laws.

Integrated management of pollution and waste are required, and environmental recovery has been made a national priority.

The role of the Maori component of the population is becoming increasingly significant for the simple reason that the Maori growth rate is considerably faster than that of the *pakeha.* Maoris comprise about twelve per cent of the New Zealand population today, and could make up as much as eighteen per cent by the turn of the century. Despite the general absence of overt discrimination, and a considerable degree of assimilation into the general population, Maoris tend to be in the lower echelon of socioeconomic development. They are increasingly concentrating in metropolitan Auckland, where their growth rate is thrice that of the *pakeha* population.

At the very bottom of the socioeconomic ladder, however, are the Pacific Islanders, immigrants from various parts of Polynesia who have come to New Zealand (very largely to Auckland) to seek their fortunes, much of which is returned as remittances to their relatives in the home islands. They face some degree of discrimination, especially from Maoris, and experience a considerable amount of unemployment and social problems. The influx continues at a high level, however, because New Zealand offers the potential of economic, educational, and recreational opportunities that far exceed those of the home islands.

New Zealand's role in world affairs has always been unobtrusive, and usually minuscule. In 1984, however, a new Labor government came to power and proudly enunciated an anti-nuclear policy that attracted considerable attention. Opinion polls show that the majority of the people supported this policy, although it was, and is, a very divisive domestic issue. The economic fallout of this development has been slight, but it is politically offensive to the governments of three countries—the United States, Britain, and France—that are important to New Zealand's trading pattern. Major changes in government took place in 1990, clouding the continuity of this policy.

In final analysis, New Zealand is a pleasant, quiet, generally happy land. Its remote location insulates it from most international problems. Although its population is still affluent by world standards, its economy now faces more difficulties than at any time in the recent past. The government has been vigorously implementing reforms, sometimes by draconian measures, and there is a continuing dismantling of the welfare state system. The final years of the 20th century will be problematic and nervous for New Zealand.

CHAPTER NINE

New Zealand's Regional Variety

Although the general characteristics of New Zealand's geography are reasonably simple and easily comprehended, the patterns of distribution are intricate and complex. Regional variation is based particularly upon topographic contrasts which profoundly affect local climatic differences. This combination, in turn, is strongly influential upon rural activities in particular and population distribution in general. Overall, New Zealand's regional pattern is much more distinctly differentiated than the gradual zonal transitions that characterize Australia.

Although only about two-thirds as large, the North Island exceeds the South Island significantly in population, agricultural output, and in commercial, industrial, and urban development.

North Island

Northland is a long, narrow peninsula that projects northwesterly for some 200 miles (320 km) from the main mass of the North Island (see Figure 9-1). It is irregularly hilly and mostly forested, and is characterized by exceedingly contrasting coastlines, smooth on the west and highly indented on the east. An abundance of sandy beaches (on both coasts) and New Zealand's mildest climate give Northland a distinct aura of subtropicality, even though its latitudinal position is equivalent to the area around Sydney in Australia.

In the early years of New Zealand's European history, Northland was a major focus of contact between Maori and *pakeha*, with significant trading and missionizing efforts. Moreover, the 1840 treaty signed by half a hundred Maori chieftains that brought New Zealand into the British Empire was promulgated at Waitangi on Northland's east coast. After that time, however,

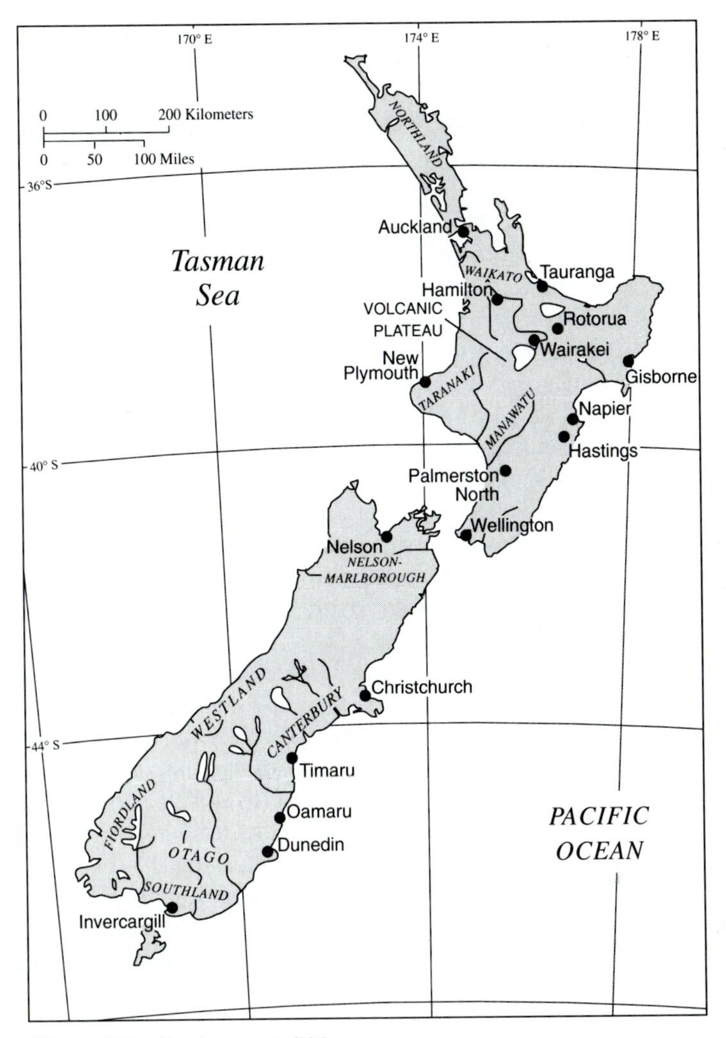

Figure 9-1 Regions and Cities

Northland became a backwater of development except for ruthless logging of its superb kauri trees for more than a century.

Today, dairying and beef cattle raising are moderately widespread in Northland, although sheep are relatively scarce. The semi-tropical climate favors the growing of specialty fruit crops, most notably citrus.

A distinctive characteristic of Northland is the relatively large Maori population; no other non-urban part of New Zealand contains such a high proportion of Maoris, despite the continuing large migration of young Maoris to Auckland.

Thanks to a warm climate, sheltered bays, sandy beaches, and proximity to the nation's largest city, Northland has recently experienced a vast increase in recreational/tourist usage, as well as continuing pressures for second home/retirement subdivisions. To develop or not to develop is a contentious issue in this area.

The proximity of Auckland, at the southern base of the peninsula, has eclipsed much urban development in Northland. The only city of note is *Whangarei* (population, 45,000), which is the location of New Zealand's only petroleum refinery, as well as other types of heavy industry.

Auckland, New Zealand's dominant metropolis, is situated at the base of the Northland peninsula on an isthmus that comprises its narrowest neck of land. There are excellent natural harbors on both sides of the isthmus, but the development of the city core was adjacent to Waitemata Harbour on the Pacific side. Despite the splendid harbors, the site had several drawbacks, including swampland, dense forests, and several dozen extinct volcanoes in close proximity. Even so, Auckland (founded in 1840) has become New Zealand's primate city, with a continuing urban sprawl, particularly toward the south. It is by far the leading industrial and trading center of the nation, and its principal population growth pole.

The *Waikato* country, immediately southeast of Auckland, is mostly a rolling lowland that encompasses much of the nation's finest farming and grazing land. The Waikato River, which flows generally northwesterly from Lake Taupo to the Tasman Sea, is New Zealand's longest.

Pakeha settlement came late to the Waikato country because of the Land Wars, and the early settlers had to exert massive efforts to clear the forests and drain the swamps. Before long, however, this became New Zealand's leading dairy area, a distinction it still holds. Intensive sheep (mostly for prime lamb) and beef cattle raising are also prominent, there is much market gardening in the northern section (near the Auckland market), and deer farming has become well established.

New Zealand's largest reserves of good quality sub-bituminous coal are in the Waikato, and are mined significantly for domestic and industrial use as well as for electricity generation. A major government endeavor over the last several years has been the establishment of a domestic iron and steel industry; this has now come to fruition based on iron sands mined from beach and dune deposits along the north Waikato coast.

Hamilton (population 110,000) is New Zealand's largest inland city and is the focal point of the Waikato district. It is a market and transportation hub, and has a growing manufacturing base.

Across some low mountain ranges to the east of the Waikato is the vast expanse of the *Bay of Plenty*, with its narrow but densely populated coastal lowland. This is another outstanding dairy area, is noted for citrus fruit, and became the principal focus of New Zealand's kiwifruit industry. The city of *Tauranga* (65,000) has become one of the country's leading ports because of expanded timber exports.

Extending south from the edge of the Bay of Plenty lowland to the shores of Lake Taupo is found what is probably the world's largest expanse of exotic timber plantations. Several species of Northern Hemisphere conifers have been introduced, but by far the largest acreage is devoted to Monterey pine.

To the general public, however, this area between the Bay of Plenty and Lake Taupo is much better known for its hydrothermal features. In the vicinity of *Rotorua* (55,000) are found most of New Zealand's geysers, hot springs, and steam vents, which comprise a carefully nurtured tourist attraction. Between Rotorua and Lake Taupo is the world's best known geothermal energy development, initiated in 1958 at Wairakei. More than 5% of the nation's electricity is now generated from natural underground steam at Wairakei.

The *central volcanic plateau*, which includes both Lake Taupo and the Rotorua area, contains a considerable variety of topography and vegetation, but a limited economic base and a sparse population. Forestry, tourism, and winter sports (primarily ski areas developed on volcanic Mt. Ruapehu) are the principal sources of income.

To the east of the plateau is a virtually continuous sequence of high hills and low mountains paralleling the eastern coast of the North Island. Between the ranges and the coast is an irregular coastal plain that is densely populated only in the vicinity of the three coastal cities (*Gisborne*, *Napier*, and *Hastings*). The hills have been deforested hastily, resulting in significant erosion problems, whereas the lowlands have been cleared and planted to exotic grasses. Extensive sheep raising is the dominant economic activity, but there are productive pockets of deciduous fruit cultivation, as well as vineyards, kiwifruit, and avocados.

West of the plateau is a broad expanse of hilly bush country that is not densely occupied by people, sheep, or cattle. Further west, however, is the remarkable *Taranaki* district. Its landscape is dominated by the beautiful, symmetrical cone of a recently extinct volcano, now officially named both Mt. Egmont and Mt. Taranaki. Streams flow down all sides of Mt. Egmont in a symmetrically radial pattern across a fringing ring-plain lowland that is one of the most attractive, mature, pastoral landscapes imaginable.

New Plymouth (50,000) a Wakefield settlement, was the early *pakeha* center, and the presence of dense forest and truculent Maoris delayed the spread of pioneers inland. Now, however, the lush green countryside of Taranaki supports one of New Zealand's densest concentrations of dairy cattle, and most of the paddocks are separated by picturesque and practical windbreaks and hedgerows of living vegetation. Onshore and, particularly, offshore deposits of recently discovered natural gas have stimulated economic development in Taranaki, especially in the regional port and service center of New Plymouth.

A narrow but productive plain extends along the coast to connect Taranaki with the fertile lowland of the *Manawatu* district. In its upper reaches the Manawatu River flows southward between two linear mountain ranges before turning westward and cutting abruptly through the ranges, from which point it shares a broad coastal lowland with several lesser streams.

Both the upper Manawatu valley and the coastal lowland have been developed (by scrub clearance, planting of exotic pasture grasses, and heavy fertilization) into some of New Zealand's most productive pastoral country. Dairying and fat lamb raising predominate; the Manawatu has the nation's densest sheep population. This is also an important area for cultivation of root crops. *Palmerston North*, New Zealand's second largest inland city (70,000), is the regional focus.

The southern prong of the North Island consists mostly of rugged hill country, with narrow linear valleys sandwiched between the ranges. The hills support extensive sheep grazing, whereas the limited lowlands have experienced considerable urbanization, focussed upon the city of *Wellington* and the two north-south transportation corridors that connect it with the rest of the North Island. Wellington was chosen as the capital of New Zealand in 1865, on the basis of its splendid harbor and its relatively central location for the country as a whole (see Photo 9-1). Its site is a difficult one for urban development, however, as there is literally no flat land. Its harborside commercial core sits largely on reclaimed land whereas the residential districts cling to the steep sides of the surrounding ridges. On the north side of the

Photo 9-1 *The central business district of New Zealand's capital city is clustered near its well-protected harbor. (TLM photo.)*

extensive harbor (called *Port Nicholson*) there has been rapid and sprawling urban development in the Hutt Valley.

South Island

New Zealand's two main islands are separated by the turbulent waters and blustery winds of Cook Strait that sometimes make the 14-mile (22-km) crossing seem a good deal longer. The South Island differs from the North in many ways—bolder landforms, higher and more massive mountains, extensive glaciation, sharper contrasts between windward and leeward locations, colder winters, less emphasis on dairying, more grain production, more modest urban development. Yet, the basic pattern of distinctive regional differentiation continues.

The northern part of the South Island, referred to by the regional name of *Nelson-Marlborough*, consists of a series of faulted ranges and valleys with a linear north-south orientation. There is much micro-climatic variety, based primarily upon topographic patterns; the western sides of the ranges receive abundant precipitation whereas the valleys have rain-shadow locations that make them some of the driest and sunniest parts of the country.

The ranges are relatively high and rugged, and are sparsely occupied by extensive sheep properties. The valleys, particularly near the coast, have relatively dense rural settlement in a small farm pattern, and there are numerous urban places of modest size. Some dairying, as well as sheep and beef cattle raising, is carried on, but the region is much better known as New Zealand's principal source of apples, pears, tobacco, and hops. Moreover, there has been considerable development of viticulture in recent years. Pine plantations and oceanic fishing add to the variety of the regional economy. Numerous bays, abundant sunshine, and attractive scenery make this area very alluring to tourists and recreationists, so that there has been considerable development of resorts and holiday homes. *Picton* is the port focus for inter-island transport, but the principal city is *Nelson* (45,000).

The mountainous spine of the South Island consists of the high country of the *Southern Alps*, a rugged and spectacularly glaciated range. Massive glaciers and highland icefields (there is a total of about 3,150 glaciers in the Southern Alps) occupy the highest parts, primarily around Mt. Cook (the country's highest peak at 12,350 feet [3,765 m]) and Mt. Aspiring (9,960 ft. or 2,990 m). The western flank descends with precipitous abruptness to the Tasman Sea. The eastern side is somewhat gentler, and its numerous glacially carved valleys are occupied by long, deep lakes and/or lengthy, braided rivers.

Below the snowline on the west side dense forests prevail. On the eastern side, however, vast expanses of tussock grassland are grazed by hardy Merino sheep, as well as various forms of exotic wildlife (particularly rabbits, deer, chamois, and tahr). Wool is the main product of the extensive pastoral "runs" of the high country, whereas on the lower slopes there is more intensive stocking of crossbred sheep (for meat and wool) on smaller holdings. Other than pastoral homesteads, permanent human habitation is scarce, but the magnificent scenery of the high country (especially around Mt. Cook, Lake Wakatipu, and Lake Wanaka) is a major attraction for both international and domestic tourists, and the abundant winter snows have made it possible to develop first-class ski areas.

In the far southwestern corner of the island, the Southern Alps are lower in elevation but even more dramatic as a scenic spectacle because many of the west-side glacial gorges of this crystalline rock massif have been flooded by the sea to form deep, steep-walled fiords (see Photo 9-2). Lengthy fresh-water lakes are common on the eastern side of the massif, and the ice-scoured uplands contain an abundance of cascading rivers and thunderous waterfalls. Remoteness, heavy precipitation, frequent winds, low winter temperatures, and an abundance of mosquitos and sandflies in summer make *Fiordland* a region that was unattractive for settlement by either Maori or *pakeha*. The area, however, is an important source of hydroelectric power, and is world famous as a tourist destination.

The western margin of the South Island is known as *Westland*. It is a narrow littoral zone in which the steep slopes come directly to the sea in

Photo 9-2 *Mitre Peak sours above the waters of Milford Sound in Fiordland. (TLM photo.)*

many places, with narrow patches of coastal plain interspersed. Abundant moisture is characteristic of this west-facing region that receives an onshore airflow most of the year. Annual precipitation exceeds 100 inches (2,500 mm) virtually everywhere, and some exposed slopes receive more than four times that amount. Dense forests, largely of southern beech, are widespread, and commercial logging is a modest but long-continuing enterprise that has considerable potential for expansion.

Agriculture is difficult under such environmental circumstances, and the population is sparse. A gold rush in the 1860s brought many people to Westland, but there have been few attractions to settlers since then. The northern part of Westland has New Zealand's largest high-quality (bituminous) coal measures, but it is a high-cost production area, due to its contorted geology and its isolation. Marginal farming and pass-through tourism provide some variety to the local economy.

The eastern side of the South Island is a total contrast to the western. An extensive foothill area borders the Southern Alps on this side, flattening out into New Zealand's most expansive lowland, the *Canterbury Plains*, a 40 × 200 mile (64 × 320 km) reach of coalescing alluvial fans composed of "shingle" (glacio-fluvial gravel) deposited by the numerous braided rivers that flow west-east from Alps to ocean. The rain shadow location, the high porosity of the shingle, and parching downslope foehn winds accentuate the dryness of the climate. Average annual precipitation varies from 25 to 35 inches (600 to 900 mm), mostly falling in summer.

The Plains were grass-covered at the time of *pakeha* settlement, and extensive sheep-raising and grain-growing were well established by the 1860s. Today the foothills are grazed by sheep in typical extensive operations, and the lowlands are completely occupied by well-developed and fairly large farms. The agrarian landscape has one of the most impressive displays of hedges (live fences of Scotch gorse) and shelterbelts (mostly of Monterey pine and Monterey cypress) to be found anywhere. A combination of fat lamb raising and grain growing dominates the rural scene, and wool is a prominent product as well.

Canterbury is the source of most New Zealand wheat and barley, as well as other cereals, and grass seed and lucerne are widely grown. Recent irrigation developments have favored the expansion of fruit and vegetable production, particularly near Christchurch, which is the focal point of the region. *Christchurch* (300,000) was founded as a planned settlement in 1840 and has long been the hub of the South Island. Its harbor (busiest in the South Island, but with a much lower level of activity than Auckland or Wellington) is in the drowned complex caldera of the volcanic mass of Banks Peninsula, just to the east of the city.

The broad Canterbury Plains shrink to a narrow coastal plain south of Timaru and pinch out completely south of Oamaru, which marks the northern margin of the *Otago* district. Inland Otago contains some of New Zealand's greatest landscape diversity, ranging from the second highest portion of the Southern Alps to the splendor of gigantic lakes Wakatipu and Wanaka to a vast expanse of tussock grass-clad hills to pleasant agricultural valleys. The hills are devoted to the extensive raising of sheep and beef cattle, whereas fat lamb raising and dairying dominate the valleys, although there is some cultivation of grains, fruits, and vegetables. *Dunedin* (110,000), which was New Zealand's largest city during the Otago gold rush days (1860s), is the dominant regional center.

The southernmost portion of the South Island is called *Southland.* Except for Fiordland, which officially is part of the Southland district, this is a region of low, rolling hills and small plains. The dairy industry of New Zealand was pioneered in Southland, and it is still the leading dairy section of the South Island. However, lamb fattening, wool growing, and beef cattle raising have become increasingly important. The port for the region is the city of *Invercargill* (55,000), which is also the hub of significant oceanic fisheries and the site of a major aluminum plant (where Australian bauxite is brought together with Fiordland hydroelectricity).

New Zealand's "third" island lies 15 miles (24 km) beyond the southern tip of South Island. *Stewart Island* contains only 675 square miles (1,750 km^2), which is less than 1% of the total area of the country. It is a hilly to mountainous island, with several peaks exceeding 2,000 feet (600 m) in height. Weather conditions are generally cool, wet, and windy, and the island is mostly forest covered. About 500 people live on the island, mostly in the northeastern corner where there is some shelter from the prevailing westerly winds. Fishing and tourism are the only economic activities of note.

CHAPTER TEN

Islands of the Pacific

Spread over the vastness of the Pacific Ocean is an uneven scattering of islands. These islands are largest and most complex toward the southwest (Melanesia), but rapidly become less diverse toward the east, northeast, and north (see Figure 10-1).

The Physical Nature of the Islands

Despite the pervading influence of the ocean, there is considerable physical variety among the islands of the Pacific. This diversity is in large measure the result of variations in relief; thus the most important differences are between the "high" islands and the "low" islands. Other significant environmental controls include island structure and lithology (especially coralline versus non-coralline), exposure (windward or leeward), and position with respect to planetary wind circulation patterns (easterlies, westerlies, monsoons).

The Geologic Fundament. The land masses of the Pacific, large and small, can be divided into four essential groups on the basis of their geologic underpinnings, primarily lithology, structure, and areal extent. Although there are notable variations within each group, meaningful generalizations can be made.

Continental Islands. A few of the lands in the region under discussion are vast enough to be divorced from simplistic environmental relationships. They are sufficiently large so that the oceanic influence is not pervasive; their terrain is sufficiently varied so that low latitude location is muted by the harsher effect of altitude; and the resulting environmental diversity renders invalid most broad generalizations. There are four such "continental islands"

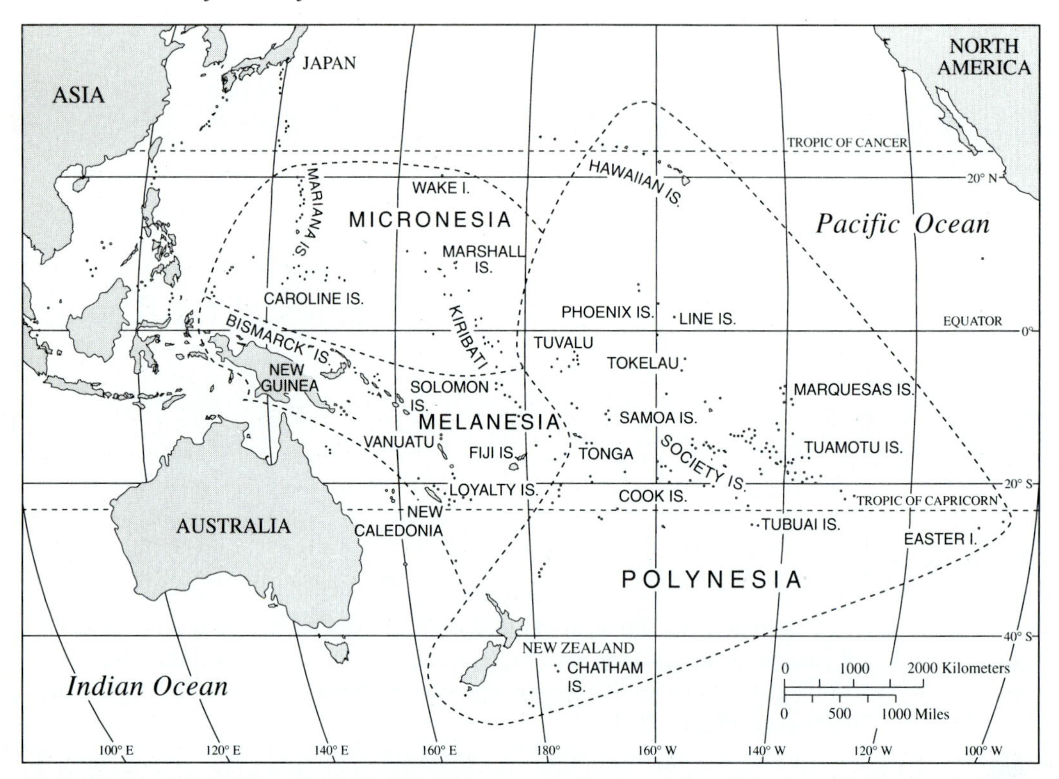

Figure 10-1 Culture Realms of the Pacific

in Australasia: Australia itself, New Guinea, and the two main islands of New Zealand. Any meaningful understanding of their characteristics requires individual analysis.

High Islands. The "high" islands of the Pacific are typically large in size, varied in relief, and steep of slope. Some are totally of volcanic origin, usually associated with the "Pacific rim of fire," that unstable oceanic perimeter that comprises the Pacific coastal margins of South America, North America, and Asia, and is distinct and conspicuous in the Melanesian portion of Australasia. Others have a less spectacular, if more complex, geologic structure, consisting of a variety of types of bedrock interspersed with igneous intrusions. A number of the volcanoes are still active, but most are extinct, or at least quiescent. Whether active or dormant, however, the influence of vulcanism on the landscape has been significant.

Some of the high islands consist of a single peak rising from the ocean floor, like Mooréa in French Polynesia; others are composed of multiple volcanic peaks, like the Bougainville; and still others comprise a range or chain of mountains, as New Caledonia. Whatever the pattern, however, almost all share certain basic characteristics. The slopes are steep and abrupt, the valleys are deep and small, the streams are swift and clear. Mountains, hills, or both make up the bulk of the island area, and coastal plains are either absent or of quite limited extent (see Photo 10-1). Local relief is often spectacular, with peaks rising thousands of feet within only a short distance of the shore. Mountains reach elevations exceeding 6,000 feet (1,800 m) on Savaii in Western Samoa, 7,000 feet (2,100 m) in Tahiti, 10,000 feet (3,000 m) on Bougainville in the Solomons, and 13,000 feet (3,900 m) in Hawaii. Even on islands where the peaks reach only 4,000 or 5,000 feet (1,200 or 1,500 m) in height, clouds persist about the summits much of the time.

Uplifted coral platforms. The occurrence of coralline structures in the Pacific is widespread. Many are in the form of narrow reefs, but some appear as more extensive platforms or amorphous masses. Some such platforms have been uplifted above the surface of the sea, forming islands of small to medium size. The tectonic uplift is sometimes complex and may involve tilting or faulting of the platform. The resulting island is typically low-lying, but its surface form may be irregular and occasionally shows considerable relief. Such islands do not have the physical diversity that characterizes volcanic islands, but most have sufficient variety of terrain to keep the coral platform from having a homogeneous environment. Examples of coral platform islands

Photo 10-1 *Tahiti is a typical high island in Polynesia. Steep slopes and luxuriant vegetation are characteristic, as are waterfalls and rushing streams. (TLM photo.)*

include Makatéa (in the Tuamotus), Nauru, Tongatapu (in Tonga), and Ocean Island.

Atolls. The most common type of island in the Pacific, by far, is the coral atoll. Atolls are remarkable entities composed of the massed and cemented calcareous skeletons of small anthozoan animals called polyps. These relatively long and narrow coralline structures are normally built in shallow water fringing some island or other land mass. Most atolls are connected in some fashion to volcanic cores of dense basalt, although in many instances the cores are deeply submerged beneath the sea. Diastrophic movements of the earth's crust often raise the coral slightly above sea level, where, in the normal workings of nature over a long period of time, soil may form, vegetation may grow, and a new island may be formed.

As the coral structure is normally begun as an island-fringing feature, even though the island that it initially fringed may subsequently sink beneath the sea, the atoll typically takes a roughly circular form, separating the active waters of the open ocean from the quiet waters of the enclosed lagoon. The term "atoll," in fact, implies a ring-shaped structure. In actuality, the ring is rarely a complete enclosure; rather it consists of a string of closely spaced coral islets (the individual islet is called a "motu"), separated by narrow water passageways.

An atoll is a typical example of a "desert island." It is low-lying (often its highest point is only a few feet above sea level) and flat, narrow and sandy, and in many cases has no surface streams at all. They are special islands that engender special feelings in people familiar with them. In the colorful prose of James Michener:

> A coral atoll, circular in form, subtended a shallow lagoon. On the outer edge giant green combers of the Pacific thundered in majestic fury. Inside, the water was blue and calm. Along the shore of the lagoon palm trees bent their towering heads as the wind directed. . . . The world contains certain patterns of beauty that impress the mind forever. . . . The list need not be long, but to be inclusive it must contain a coral atoll with its placid lagoon, the terrifyingly brilliant sands and the outer reef shooting great spires of spindrift a hundred feet into the air. Such a sight is one of the incomparable visual images of the world.[1]

Although the term "atoll" was actually coined to apply to certain islands in the Indian Ocean, the great majority of the world's atolls are in the Pacific. Nearly all of the Micronesian islands, most of the Polynesian islands, and a large share of the Melanesian islands are atolls.

[1] James A. Michener, *Return to Paradise* (New York: Random House, Inc., 1951), pp. 6–7.

The Fringing Reefs. Regardless of type, practically every tropical island in the Pacific is bordered by, or enclosed by, a coral reef.[2] The largest coralline feature in the world is Australia's Great Barrier Reef. Of almost similar magnitude is the fringing reef that encircles the island of New Caledonia. The ubiquity of reefs associated with the Pacific islands is a reflection of the extraordinary abundance of the fragile but fecund coral polyp. They occur in uncountable billions in the tropical waters of the world, and the structures they construct are massive and sturdy. Accordingly, their contribution to life in the Pacific Basin is incalculable, and their presence in the oceanic landscape is immensely important (see Photo 10-2).

Floristic Patterns. The Pacific islands house a remarkable diversity of plant communities, although, for the most part, the number of species is relatively few (except in New Guinea), doubtless due to isolation. A further function of isolation is the large proportion of endemic plants (those found only on one or a small group of islands, and nowhere else in the world). For example, more than 70% of the native floral species of both New Caledonia and Hawaii are endemic. The heavy rainfall of the high islands, combined with the characteristic high temperature and humidity of the tropics, produces an environment that supports a perennially green, almost overpoweringly luxuriant vegetation. The dense vegetation of the rainforest, with its tall trees, multiple canopies, abundance of parasitic growth, and sunless floor, is most widespread in Melanesia, where the extensive mountains of such large islands as New Guinea, New Britain, Bougainville, Guadalcanal, and Espíritu Santo intercept a great deal of precipitation. Smaller areas of rainforest are found on other islands, such as the windward slopes of Viti Levu in Fiji or certain southeastern hillsides on Tutuila in American Samoa, but it is clear that only an optimum combination of elevation and exposure can produce enough rainfall to support a rainforest association.

All is not rainforest in the high islands. Many areas have less luxuriant natural vegetation, due to lower annual rainfall totals or to seasonal dry periods, or both. More open forests occur in many localities, savanna woodlands are found on leeward slopes and protected exposures, and swamp associations occupy many river valleys and coastal plains. Furthermore, fairly extensive grasslands, often human-induced, have developed on some of the larger islands. Grasslands are most notable at middle elevations (4,000 to 6,000 feet or 1,200 to 1,800 m) in New Guinea, where they support the greatest rural population densities of the island. Another special floristic association in New Guinea is the moss or cloud forest, which is characterized by stunted trees, thick mossy ground cover, and almost continual mist. It occupies fairly extensive areas between about 7,000 feet (2,100 m) and 11,000 feet (3,300 m) elevation.

[2]Some islands in the eastern Pacific, like the Marquesas, have very restricted reef development, apparently owing to the upwelling of cooler water that is unsuitable for the survival of coral polyps.

Photo 10-2 *A low air view of a portion of the coast of the island of Viti Levu in Fiji. The fringing reef separates the waters of the open Pacific from the coastal lagoon. Two breaks can be seen in the reef; they are caused by a discharge of fresh water from streams on the island. Corals cannot survive in water of low salinity. (TLM photo.)*

The coral islands are so small and low that their vegetation is much sparser. The drier conditions, occasioned by scanty rainfall and porous soil, support a relatively uniform flora of maritime and drought-resistant plants. Apart from the ubiquitous pandanus and coconut palm, and various plants cultivated by the islanders, there is often little more than scattered grasses, bushes, and herbaceous plants. Many of the smaller islands are entirely treeless, and the number of native plant species on an atoll is often less than two dozen.

Of all the plants of the Pacific Basin, none is more notable or significant than the coconut palm. It is the conspicuous species of the coastal lowlands; the island whose beaches are not fringed with palm trees is a rarity indeed. The extraordinarily widespread occurrence of the coconut palm apparently represents a combination of natural diffusion (the coconut is capable of floating some distance over the ocean, and then lodging on a beach and sprouting) and cultural diffusion (its usefulness is legion, so that Pacific islanders frequently took coconuts with them on their migrations from one island to another). Almost as common, but not nearly as useful as the coconut, is the pandanus or "screw pine" (*Pandanus* spp.), the seeds of which also float.

Fauna of the Islands. There is considerable diversity in the faunal complement of the Pacific islands, as might be expected with habitats ranging from mountainous rainforests to barren, sandy atolls. Overall, however, animal life is notable for its sparsity, particularly with reference to terrestrial forms. The barrier effect of a great ocean severely restricts natural diffusion, so that on most islands the more conspicuous elements of the fauna are species that have been introduced by humans.

The greatest abundance and variety of terrestrial animal life is found in New Guinea, with a general decrease in numbers and variety eastward and northward from there. This is logical, considering the apparent spread of most animal life across the Pacific from Southeast Asia. Thus Melanesia has a richer fauna than either Polynesia or Micronesia, and in general the larger islands are more bountifully endowed with animal life than the smaller ones.

Even so, terrestrial mammals are quite limited everywhere. New Guinea has considerable variety, ranging from marsupials and monotremes through rodents to wild dogs and wild pigs; but even there no native primates or ungulates, and very few carnivores, are to be found. On the other islands native terrestrial mammals are very poorly represented and in many cases totally absent.

Humans have altered the faunal pattern considerably by introducing exotic animals to the islands. Rats, for example, are now almost ubiquitous throughout the Pacific Basin and serve as a major destroyer of coconuts on many islands. Feral livestock (domesticated animals that have reverted to a wild existence) are also widespread. Feral goats, sheep, pigs, and cattle are prominent on most of the Hawaiian islands, and feral pigs and goats are found in a number of other localities. Exotic wildlife species from other parts of the world also have been brought to many Pacific islands.

The avian fauna is much richer as an ocean is less of a barrier to birds than to mammals. New Guinea is world-renowned for its remarkably varied bird life, most notably represented by a multitude of species of birds-of-paradise. The other large Melanesian islands also have considerable variety in their birds, and even small atolls are likely to be well endowed. Again, people have played an important role by introducing various species, particularly the Indian mynah, the bulbul, the English sparrow, and the starling, all of which are widespread (and generally considered to be pests) throughout the Pacific.

Reptiles and amphibians are fairly common in the larger Melanesian islands, particularly New Guinea, but are much less common elsewhere. Insects and other arthropods occur in varying quantity and diversity, once again with the greatest representations in Melanesia. Crabs and other littoral crustaceans are especially numerous throughout the region.

The Peopling of the Pacific Islands

The widespread islands of the Pacific Basin have always been far removed from the mainstream of world consciousness and interest; thus, with limited exceptions, they have not been subjected to a series of major colonizations, invasions, and wars. Even so, the present population of the region is an imperfect amalgam of many diverse groups, who have spread sporadically through the islands, leaving imprints both ephemeral and lasting.

The peopling of the Pacific islands was accomplished over an exceedingly long period of time, involving numerous groups of migrants who moved out generally but erratically from Southeast Asia across the vast ocean, possibly starting as much as 50,000 years ago. The migratory sequences were varied, complex, and overlapping, and still are understood very imperfectly.

It is believed that Negritos and Australoids were the first to penetrate the Pacific, at least as far as the Philippines, New Guinea, and Australia. These were dark-skinned, small-statured, hunters and gatherers who had a low level of material culture. Their technology was primitive and their imprint on the region was small. They were generally inundated and absorbed by later, more advanced groups. Still, some Negritos exist today in New Guinea, and as late as 1967 government patrols encountered tribes who had never before seen either white people or government representatives.

Later migrants mostly arrived after 5,000 BC. Many were essentially Negroid in racial characteristics, whereas others were predominantly Caucasoid and still others possessed distinctively Mongoloid features. Racial mixing, then, has taken place on a grand scale over a long period of time, and the combination of hybridization and migration has resulted in a veritable patchwork of racial types across the Pacific. Distinctions are often made among Melanesians, Polynesians, and Micronesians, but any broad generalizations about their ethnic characteristics must be examined with great care.

The various natives were resident in the islands for hundreds or thousands of years before the first Europeans appeared on the scene. For the most part they evolved a communal village life, often extending the concept of "family" beyond blood relationship. They developed a subsistence economy based on gardening and fishing, which are still the basic forms of livelihood for many of the people in the region. Root crops, such as taro and yams, and tree crops, such as breadfruit and coconuts, were then and are now staples in their agriculture. All of the staple food crops of the region, with one exception (the sweet potato, a native of South America), were brought from Southeast Asia.

Penetration by Europeans into the Pacific was slow and sporadic. From the early 1500s, when the first explorations were made, some three centuries elapsed before the European powers began to express serious interest in colonization of the islands. The first century of European exploration (the 16th) was primarily an Iberian century, with Portuguese and Spanish ships in the majority. The Portuguese captain Abreu was the first definitely recorded European visitor; his exploration was mostly restricted to coastal New

Focus Box: A Polynesian Settlement Model

Polynesia was just about the last part of the habitable world to be settled by humans; occupance dates from the last 1,500 years or so. The settlers were basically homogenous from an ethnic and cultural standpoint, so the patterns that developed illustrate a controlled laboratory-type situation.

High Islands. The typical high island in Polynesia consists of a steep-sided central peak with ridges and valleys extending alternately and radially outward and downward to the coastline. Permanent streams flow down each of the valleys to the sea. A placid lagoon surrounds the island, its outer edge marked by a coral reef through which there are several tempestuous, natural passages.

The drainage basin is the basic environmental unit, and the people of each watershed were likely to be governed by a different chief. Subchiefs might exercise control of parcels within the watershed, but the habitat was divided so that each sub-group had coastal land, stream land, forest land, and gardening land.

Chiefs might attempt to extend their area of authority, but a power struggle to control an entire island was difficult because the island was relatively uniform on all sides, and the resources (e.g., stone, wood, bone, fibers) were more-or-less equally available in each valley. Thus, no monopoly of resources was likely.

Atolls. Essentially the same people arrived as settlers to the atolls as to the high islands. However, the atoll environment was quite different, and the pattern of resulting occupance was clearly distinct. The land area of the atoll is low-lying, and lacking in water and luxuriant vegetation. Its resources are few; there is an absence of stone (except for coral) and only a limited range of timbers and fibers. These sparse resources tend to be distributed relatively uniformly.

Once the atoll population exceeded a couple of hundred, the people began to divide into political groups, their settlements invariably focussing on passages in the reef. Atoll society tended to be less structured than that of high islands. Moreover, the political and social units normally were considerably smaller. Such population limiting factors as war, infanticide, and abortion became significant fairly early on atolls.

Arrival of Europeans. With the arrival of exotic and powerful intruders, native society was rapidly shaken to its roots. The introduction of new technological items (metal tools, guns, utensils) caused the devaluation of many old "resources," such as stone and fibers. New food crops were introduced, with new tools to cultivate them.

Shortly after the introduction of muskets, one chief became dominant on each island. Who got the muskets? Which chief was favored? European ships were likely to go to the lee side of the island and seek the best reef passage. This passage became the European "port" for the island. The Europeans dealt directly with the chief of the local watershed or village, which immediately strengthened his power, and he usually became the high chief of the island. Eventually he would expand his control to neighboring islands, which resulted in the first Polynesian kings.

Europeans were less interested in atolls because they had less to offer. Thus atolls usually became dominated by high islands, and often suffered rapid depopulation.

There are many exceptions to the generalizations of this settlement model, but its broad validity is demonstrated in the history of Fiji, Tonga, the Cook Islands, and some of the Society Islands.

Guinea. Magellan's Spanish crew made the first circumnavigation of the globe in the early 1520s, but managed to see less than half a dozen islands. Various other Spanish and Portuguese ships navigated the Pacific during that century, but their records were singularly poor, and many of their discoveries apparently went unrecorded. During the 17th century, Dutch explorers were in the ascendancy. The Netherlands established its hegemony in the East Indies, and Dutch navigators traveled widely in the Pacific. The 18th century was a time of increasing exploration in the Pacific Basin for ships from many nations; however, the French and British were dominant. The greatest of all Pacific explorers, Captain James Cook, made his three voyages during the 1760s and 1770s.

Following soon after the explorers, the islands were intruded upon by European gatherers and hunters (of sandalwood, pearl shell, whales, and other specialized products) and by traders bartering European goods. Missionaries also made a significant imprint on social and political life, and stimulated production for trade. Some Europeans had begun to settle on Pacific islands as early as the 16th century; however, the general region did not come under significant European influence until after 1800. From then on, native culture and institutions, as well as the people themselves, began to disappear or become significantly modified with appalling and growing rapidity. By the time the 19th century ended, most of the native communities into which Europeanization had penetrated had dwindled greatly in numbers. Furthermore, on a number of islands a new demographic element was introduced when Asian laborers came to work on plantations and in mines.

As late as 1840, practically all of the Pacific islands were unclaimed by European powers, with the exception of Spain's requisition of some of the Marianas. After that time, however, colonial powers (primarily Britain, France, Germany, and the United States), motivated by a variety of interests, jockeyed for position. There were several confrontations of naval strength, particularly focused upon Samoa. Often it seemed that a military crisis was about to erupt in the Pacific, but events such as the Franco-Prussian War, the Boer War, and the Apia (Samoa) hurricane disaster of 1889 distracted the participants.

By the turn of the century, European (including American) powers had taken over the Pacific Basin. Germany entered the 1900s as an important colonial power; she controlled Western Samoa, Northeastern New Guinea, the Bismarck archipelago, Nauru, and the principal islands of Micronesia.

Britain governed Papua (southeastern New Guinea), the Solomon Islands, Fiji, and a number of minor island groups, and had a protectorate relationship with Tonga. France ruled New Caledonia, the far-flung islands of French Polynesia, and (with Britain) the condominium of the New Hebrides. The principal possessions of the United States were Hawaii, Guam, and eastern Samoa.

Germany was stripped of her overseas possessions during World War I, and Japan made an appearance in the Pacific island scene at this time when German Micronesia was given to her in mandate. Western Samoa was mandated to New Zealand, northwestern New Guinea and the Bismarcks went to Britain and Australia jointly, and Nauru was made a cumbersome triple mandate of Britain, New Zealand, and Australia.

World War II brought abrupt and lasting changes to the Pacific islands. Moving from secret bastions in Micronesia (which had been closed to the outside world since the mid-1930s) and her invading spearheads in the East Indies and the Philippines, Japanese military might quickly overran about half of Melanesia. A lengthy series of bloody battles was fought in New Guinea and the Solomons, and air and sea attacks spread devastation in the Bismarck archipelago and throughout Micronesia. Although most of Polynesia was spared direct battle damage, it shared with Melanesia and Micronesia the significant economic and social repercussions that go with three-and-a-half years of adjacency to major region of warfare. Stone-Age Melanesians, Micronesians, and Polynesians were jerked abruptly into the international realities of the atomic age, and the South Pacific has never been the same since. Native standards and values were rapidly abandoned or significantly modified, a cash economy did away with the last vestiges of a barter system, material possessions from the Western world achieved great desirability, transportation and communication networks were vastly improved. For the first time, solid feelings about colonialism and nationalism began to develop, though at a slower rate and on a smaller scale than in most other parts of the underdeveloped world.

Economically, the Pacific islands enjoyed a general period of prosperity in the immediate post-war years. Private enterprise was officially discouraged at first, largely because the major colonial powers had socialist-trending governments in power. The pendulum soon swung the other way, however, and private investment was strongly encouraged. For a few years after the war most of the islands reaped the benefits of high prices for their primary produce (copra, cocoa, coffee, sugar, and rubber) but later the price structures leveled out, and virtually all of the island groups reinforced their prewar economic position of being financial drains upon the colonial powers.

In ensuing decades, political, social, and economic changes have come with great rapidity. Starting with Western Samoa in 1962, most of the island groups have achieved independence or a strong measure of self-government. Only in the French colonies are nationalistic fervors still unsatisfied. Social stratification and ethnic discrimination have been diminished. Despite varying degrees of diversification, the economy of the islands still must contend with the triple handicaps of limited resources, lack of capital, and remoteness from world markets.

As the 20th century approaches its conclusion, perhaps the most far-reaching continuing development in the Pacific Basin is the improvement and expansion of transportation patterns and the concomitant accelerated growth of tourism. Heralded by the achievement of statehood by Hawaii in 1959 and the building of a jet runway across Tahiti's northwestern lagoon in 1960, the world has discovered the Pacific. Tourist travel to Hawaii grew from 185,000 in 1958 to 7,000,000 in 1992, and there has been similar proportional growth in Tahiti, Fiji, Guam, and other places. Indeed, such previously inaccessible localities as Bora-Bora, Rarotonga, and Western Samoa are on the verge of tourist inundation.

Contemporaty Ethnic Diversity

The present population of the Pacific islands (not including Australia, New Zealand, or the Indonesian portion of New Guinea) is about 5.5 million. Roughly 80% of this number represents indigenous people, and of the remaining 20%, about two-thirds are Asian and one-third European. Of the indigenous population, more than 80% is Melanesian (including Papuan), and the remainder is about two-thirds Polynesian and one-third Micronesian.

The population of Fiji represents a unique case in the Pacific. Until very recently its largest population component has consisted of ethnic Indians, who are descendants of indentured laborers imported to work in the sugar cane fields two or three generations ago. Since about 1990, however, there has been considerable emigration of Indians, due to two Fijian-led military coups, with the result that Ethnic F ijians now comprise a slight majority of the population (see Photo 10-3).

The situation in New Caledonia is different still. Its population of about 175,000 has approximately equal cohorts of Melanesians (*kanaks*, as they call themselves) and Europeans (predominantly French), which together comprise about four-fifths of the total population. The remainder consists largely of Polynesians and Asians, most of whom are either Vietnamese or Javanese descended from indentured workers imported in the past.

Guam's population of some 125,000 is nearly half *Chamorro* (the native Guamanian, which is a mixture of Micronesian, Polynesian, Filipino, and Spanish blood), 25% North American, and 22% Filipino.

Apart from the three situations just described, non-native peoples comprise only a small minority of the population. Europeans (a term that includes North Americans, Australians, and New Zealanders) are found in limited numbers, primarily in the larger towns of such islands as Tahiti, American Samoa, and Papua New Guinea. Asians occur in a more disseminated pattern. Chinese, for example, are a small but significant minority on many islands, especially in French Polynesia. Vietnamese are also notable

Photo 10-3 *Ethnic variety in Fiji. This scene from the central market of Suva highlights the contrast between Fijian and Indian. (TLM photo.)*

on many French islands. Many Japanese migrated to Micronesia during the 1920s and 1930s, but most have now returned to their homeland, and only a scattered few remain.

Linguistic variety is even greater than ethnic diversity in the Pacific islands. Most indigenous peoples of the region have as a native tongue some Austronesian language, which are prevalent from Madagascar in the western Indian Ocean to eastern Polynesia. The principal exception is the 700-or-so Papuan languages of Melanesia, most of which are found in Papua; these do not comprise a distinct linguistic family, rather a miscellany that are often totally unrelated.

Melanesia has an incredible linguistic complexity. Nearly one-fifth of all the living languages in the world are endemic to Papua New Guinea, where there is an average of one language for each 5,000 people, and most of them are mutually unintelligible. In Vanuatu, the linguistic fragmentation is even greater; there is a ratio of one language for every 1,250 people.

In Polynesia and Micronesia, on the other hand, there are relatively few languages, and adjacent languages tend to be more-or-less mutually intelligible (e.g., Samoans can understand Tongan) European languages, primarily English and French, are used to a greater or lesser extent in all of the islands, and are considered to be official languages in many of them. The only other *lingua franca* of significance is *pidgin*, which is used fairly widely in Melanesia, but not elsewhere in the region. It comprises an unruly jargon of European (especially English) and indigenous words, using native syntax. *Pidgin* is especially important in Papua New Guinea, where it is recognized as an official language.

Religious convictions are strong in most of the Pacific islands. Animism is still prominent in some remoter localities, and Asian religions (Hinduism, Islam, Buddhism, etc.) are notable in areas of Asian settlement. Most islanders, however, are Christianized, with a variety of Protestant denominations (especially Methodism and Mormonism) and Roman Catholicism being particularly notable. Church buildings are numerous and often the most prominent structures in the towns and villages, particularly in Polynesia.

Population densities vary significantly from island to island, following no broadly predictable pattern. In general, the greatest densities occur in groups of small islands, such as the Tuamotus or the Ellices. New Guinea, despite its large total population, has a relatively low population density.

The great majority of the peoples of the Pacific islands live in a rural environment, normally in small agricultural or fishing villages. These hamlets are often discrete entities, functionally remote from neighboring villages that may be only a very few miles distant. Frequently the cultural patterns vary significantly from village to village, and sometimes mutually incomprehensible dialects are spoken in neighboring valleys.

In recent years, there has been a prominent migration of people from rural areas to urban centers, from outer islands to main islands, and (on the high islands) from highland regions to coastal districts. As a result, the larger towns are experiencing explosive population growth (from high birth rates as well as inmigration) that is producing social and economic problems that are overwhelming for the relatively impoverished governments. (A former Prime Minister of Papua New Guinea noted, "In all the 700 languages of our country we have never needed words for slums, for unemployment, for air pollution; but now we do.")

Apart from Honolulu, with a population exceeding 850,000, there is no real metropolitan center in the Pacific islands. However, four other urban places (Port Moresby in Papua New Guinea, Suva in Fiji, Noumea in New Caledonia, and Papeete in Tahiti) now have populations of more than 50,000. These, along with many other towns, continue to grow rapidly.

Emigration serves as an important "safety valve" to ameliorate overpopulation in Polynesia, although it is a relatively minor phenomenon in Micronesia and Melanesia. There are two dominant flows—southward to New Zealand, and northward to North America. Auckland, New Zealand's capital city, now contains more Niueans than Niue, more Tokelauns than Tokelau, more Cook Islanders than the Cook Islands, and about one-eighth of all the world's speakers of Samoan. Another one-eighth of all Samoan-speakers live in the United States, primarily in Hawaii and California, and there is a prominent flow of Tongans to the United States as well. Indians from Fiji are migrating in increasing numbers to Canada.

The Limited Options of Island Economies

A major continuing problem on almost every Pacific island is the development and maintenance of a viable economy. Resources are generally few, technology is usually basic, population pressure is normally great, capital is customarily limited, and external markets are distant and uncertain. Farming and fishing are the basic economic activities, both at subsistence and commercial levels. In addition, there are a few prominent nodes of mining, and commercial forestry has been developed on some of the larger islands. For many of the island nations, however, the twin economic keystones are foreign aid supplied by "European" countries, and tourism.

At the most basic level there is still a very important role in human nutrition played by subsistence hunting/gathering/fishing activities. Subsistence fishing is widespread in the Pacific, and hunting/gathering is notable throughout Melanesia. The largest total dietary contribution probably comes from the wild sago palm, whose starchy center is gathered extensively in the lowlands of Papua New Guinea and to a lesser extent in other parts of Melanesia. Wild yams, bananas, breadfruit, pandanus, and a wide variety of other fruits, nuts, and greenery are collected for subsistence consumption. Almost all forms of animal life, from crocodiles and wild pigs to moths and beetle larvae, are sought for their protein value. And of inestimable importance for coastal dwellers is the variety of marine life that can be collected at low tide through systematic reef gathering.

Subsistence farming in the Pacific isles is a traditional activity that focusses on small "garden" plots associated with village settlements. Technologically speaking, most such agricultural systems are of the shifting type, generally referred to as *swidden* or slash-and-burn, in which the native vegetation is roughly cleared and burned, and crops are planted for several years before the original plot is allowed to revert to the native flora again and a new plot is cleared. The type of shifting cultivation varies from island to island, with yams, taro, sweet potatoes, bananas, and cassava being the major staple crops.

Although shifting cultivation predominates on most of the high islands, more intensive permanent systems of subsistence farming are also common, particularly on the low islands. Many of them focus upon irrigated or swamp patches of taro and giant taro. Whether shifting or permanent, hand tools, ranging from digging sticks to steel spades, predominate. There is very little use of fertilizers or chemical pest controls, although intensive mulching is fairly common.

During the colonial period, commercial agriculture usually involved large-scale plantations in which European capital, entrepreneurship, and management were combined with native (or imported indentured) workers to grow tropical produce for overseas markets. There are still vestiges of this colonial system, but for the most part the large plantations have been either closed down or subdivided. They have been replaced in part by smallholder farming, in which an indigenous family grows one or more crops on a modest scale for commercial purposes.

The cropping pattern of commercial agriculture varies from island to island and from district to district. Some crops are grown in highly specialized locations (as sugar cane on the leeward side of the larger Fijian islands), whereas others are virtually ubiquitous in distribution. The coconut palm is the most widespread of the commercial crops, being cultivated by smallholders on virtually every island and grown in large plantations in many places. Bananas are also widely cultivated. Other tropical commercial crops that have more limited distribution include coffee, tea, rubber, oil palm, vanilla, pineapple, and papaya. In some places the spectrum of commercial crops has been expanded to include a variety of fruits and vegetables that were not traditional plantation crops, most notably watermelons and citrus fruit. In addition, there is (usually small-scale) commercial production of such tropical food crops as rice, cassava, sago, taro, yams, sweet potato, maize (corn), and kava.

Livestock are widely distributed in the islands, but generally occur in small numbers and are mostly kept for subsistence purposes. Pigs are the most common animals, being particularly notable in Melanesia but found throughout Polynesia and Micronesia as well. Cattle are kept in relatively small numbers, although there are commercial beef cattle enterprises of note in Papua New Guinea, New Caledonia, Fiji, and Guam.

Commercial forestry is limited to the larger islands of Melanesia, and is concentrated almost exclusively on relatively valuable tropical hardwoods. The principal developments are in Papua New Guinea, the Solomon Islands, and Fiji.

Mining in the islands involves scattered nodes of high-volume output, for the most part. Once again, the major developments are in the larger islands, with a few exceptions. Most outstanding are the multi-mineral Ok Tedi mine in the western highlands of Papua New Guinea, several gold mines in Papua New Guinea and Fiji, a couple of nickel mines in New Caledonia, and the island of Nauru, which has become a vast phosphate quarry.

Commercial fishing in the Pacific islands is a dynamic industry, with pronounced political and social ramifications. Small-scale commercial fishing has been carried out for decades on many of the islands, but large-scale enterprises have been relatively few and mostly have involved companies and boats based around the rim of the ocean, particularly in Japan, the United States, Korea and China. All Pacific island countries have now embraced the principle of a 200-mile economic zone radius beyond their shores, and they charge sizable fees for foreign fishing boats to operate within these limits. For some small, impoverished countries (such as Kiribati and Tuvalu) these

Focus Box: The Mighty Coconut

There are more than 1,500 kinds of palms, all of which are native of the tropics or subtropics. The one that is most important to humans, because of its many uses and its widespread distribution, is the coconut palm, *Cocos nucifera*. It is endemic to Melanesia and Southeast Asia, but has been introduced to all the tropical and subtropical coasts of the world.

The coconut palm is a tall (mature height of 40 to 100 feet, or 12 to 30 meters), graceful tree that is characterized by large featherlike leaves that spread from the top of its branchless trunk. Coconuts, the fruit of the tree, grow among the leaves, with a healthy tree producing 100 or so a year. Each coconut has a smooth rind within which there is a tough, fibrous husk that is 1 to 2 inches (2.5 to 5 cm) thick. Inside the husk is a hard, thin shell that encloses a layer of white, sweet-tasting "meat." The hollow center of the coconut contains a clear sugary liquid that is referred to variously as "water," "milk," and "wine" in various languages.

The dried meat of the coconut is called *copra*, and copra production is the oldest and most valuable of the basic industries of the Pacific islands. The meat is extracted from the coconut in small sections, and laid out on racks to dry (see Photo 10-4). Sun-drying produces the best quality copra, but continuous sunshine is uncertain in the tropics and rain is ruinous to drying copra, so in most places the drying process is completed in covered racks or in ovens. Most dried copra is transported to major industrial cities for crushing, where the oil (comprising some 70% of the weight of the copra) is extracted. Coconut oil is used widely in the manufacture of soap, margarine, cooking oil, and similar products.

Most tropical countries produce copra, so the international competition

Photo 10-4 *A Tahitian farmer looking over his rack of drying copra. (TLM photo.)*

is fierce, and prices gyrate widely from year to year. Southeast Asian countries—particularly the Philippines, Indonesia, and Sri Lanka—are the principal producers. Their output is exceedingly greater than that of the leading Pacific Island producers—Papua New Guinea, the Solomon Islands, and Vanuatu. However, the copra harvest is critical to almost every Pacific island country, comprising the sole source of income for many families.

Although commercial copra production is integral to most Pacific island economies, the subsistence value of the coconut palm is even greater. The trees grow plentifully and vigorously along almost every lowland coast, and the island people have learned to use them in manifold ways. The meat is used in the preparation of a great variety of dishes, and the coconut water is not only a delicious drink but is so pure that it can be used medically as a completely sterile saline solution. Moreover, both the coconut water and sap from the tree's blossoms can be used to make sugar, vinegar, and mildly alcoholic beverages.

The trunk of the tree supplies wood for the building of houses, boats, fences, and furniture. The whole leaves are used to make thatch roofs and partitions, and plaited strips of leaf can be converted into baskets, hats, mats, carpeting, and decorations. The husk of the coconut is made into yarn, rope, mats, and brushes. The coconut shell furnishes the raw material for a variety of household utensils, and makes good charcoal. Even the roots are used in the preparation of medicines, beverages, and dyes. It is little wonder that the coconut palm is so highly valued in the tropics, in general, and the Pacific islands, in particular.

fees comprise a significant proportion of the national budget. Apart from the foreign fishing fees, commercial fishing contributes relatively little to the island economies except in Papua New Guinea, the Solomon Islands, and particularly American Samoa (where two large tuna canneries have operated for many years).

In recent years, most of the islands have looked on tourism as a mystical panacea that will balance their budgets and solve their economic problems. For some, tourism has indeed brought in a flood of foreign money. Hawaii is the classic example, but tourism has also become big business in Tahiti, Fiji, Guam, and both American and Western Samoa. As hotels proliferate and airline routes are extended, the tentacles of tourism have reached out to almost every island group; there are still many small and remote islands that are untouched, but their number diminishes year by year. The affected islands soon learn, however, that tourism a mixed blessing. New money brings with it new problems, particularly in terms of breaking down conventional value systems, introducing new social problems (such as crime and substance abuse), encouraging unrealistic aspirations, and destabilizing traditional politico-economic systems.

The Islands in Brief

Melanesia. The islands of Melanesia occupy the southwestern part of the Pacific basin, generally north and northeast of Australia (see Figure 10-2).

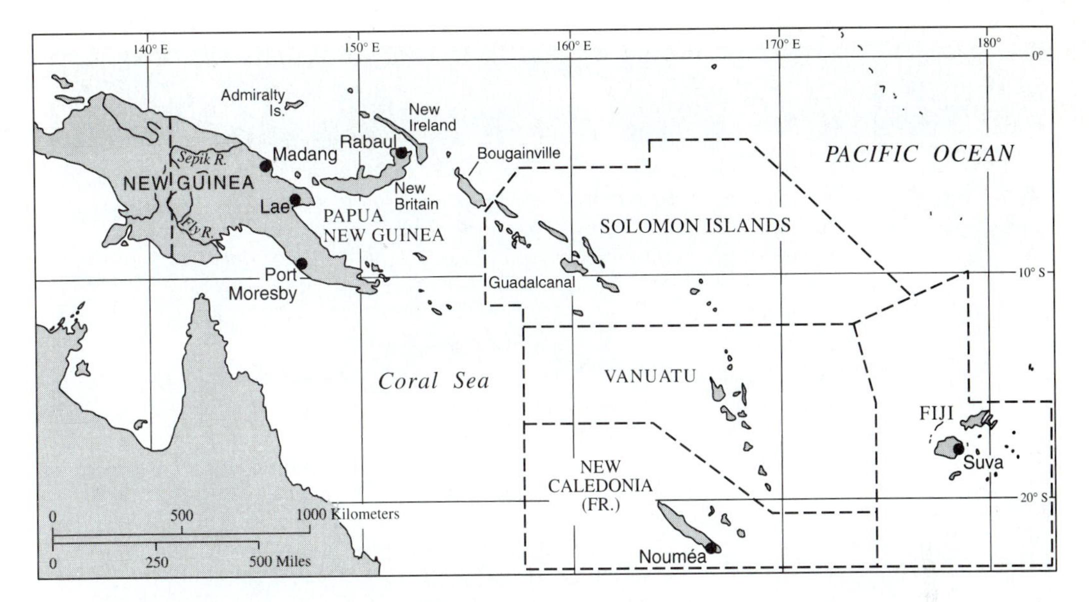

Figure 10-2 Melanesia

Although many small and low-lying islands are included, Melanesia is notably a region of relatively large, high islands. Melanesia contains more than 98% of the total land area of all Pacific islands and about 82% of all Pacific island population (excluding Hawaii). There is much cultural diversity, but most Melanesians have broadly similar physical features—dark skin, frizzy hair, and stocky build.

Papua New Guinea is by far the largest, most populous, and most diverse nation in the Pacific (see Photo 10-5). It consists of the eastern half of the huge island of New Guinea, the Bismarck Archipelago (whose largest islands are New Britain, New Ireland, and Manus), the northernmost Solomon Islands of Bougainville and Buka, and several smaller island groups just east of New Guinea. These rugged, rainy, equatorial lands comprise the home for nearly four million culturally varied people, some 98% of whom are Melanesian. Their economy runs the gamut from primitive, wandering Negrito hunters to sophisticated, urban-dwelling office workers; their social organization varies from fragmented, patrilineal societies in the New Guinea highlands to tightly organized and densely settled matrilineal groups in the coastlands of New Britain. This is one of the most complicated linguistic areas in the world, with more than 700 languages and dialects spoken, and no individual language spoken by more than about five per cent of the population, although the native pidgin is understood by perhaps one-third of the people.

Nearly half the total population resides in small villages and practices a shifting garden type of subsistence farming. Their principal subsistence crops are sweet potatoes (by far), taro, bananas, yams, and sago. The densest population concentrations are in the "highlands", relatively fertile upland

Photo 10-5 *A great variety of produce is displayed in this portion of the huge Koki market in Port Moresby, Papua New Guinea. (TLM photo.)*

valleys at elevations of 4,000 to 6,000 feet (1,200 to 1,800 m) in the interior of the main island.

Although the vast majority of the people still live in villages, in recent years there has been a very high rate of urbanization. Each of the dozen or so largest towns has attracted vast squatter populations, most of whom are unemployed. This has brought about enormous social dislocation, as manifested principally in an astounding crime rate (particularly robbery and assault). The population of Port Moresby has skyrocketed to more than 200,000, making it the largest urban place south of Honolulu, north of Australia, and east of Indonesia.

Since independence in 1975, the national economy has been supported largely by foreign aid, particularly from Australia, which was the colonial power that granted independence. Mining has been the principal revenue-generating industry. It was dominated by the gigantic Panguna copper enterprise on Bougainville until the mine was closed by secessionist revolutionary activity in 1989. Major mining ventures currently operating include the enormous Ok Tedi gold/silver/copper project in the western highlands, the Porgera and Lihir gold mines, and the Kutubu oil field.

There is a growing component of commercial agriculture, primarily copra, palm oil, cocoa, and coffee. Cattle raising is increasing, as is commercial fishing, logging, and tourism. However, Papua New Guinea faces enormous economic handicaps for the foreseeable future. The people, for the most part, are unsophisticated and uneducated. The country is fragmented, by topographic impediments, cultural schisms, political rivalries, regional antagonisms, and linguistic barriers. Transportation is extraordinarily difficult; there is no railway, limited mileage of useful roads, and only a single good port.

Despite this array of handicaps, Papua New Guinea has established a viable democratic government within a single decade, and is a highly respected country in the small world of the South Pacific. Its challenges are enormous, but it has already come further than many observers predicted in the recent past.

The *Solomon Islands* consists of 10 high islands and many small ones extending in a double chain for nearly 1,000 miles (1,600 km) southeast of Bougainville. The islands are physically similar to those of the Bismarck Archipelago, and the indigenous economy is much like that of Papua New Guinea, with village-oriented subsistence agriculture as the principal activity. The people, however, are not nearly as varied, either physically or culturally, as the New Guineans, being of basic Melanesian stock and having fewer languages. The total population of the country is approaching 300,000, of which 95% are Melanesian.

Commercial fishing has emerged as the bulwark of the national economy. It is based largely on tuna, mostly for export to Japan, and is significantly supported by Japanese capital and management. Forestry is the second export earner, mostly involving the export of hardwood logs to Japan. Commercial agriculture revolves around copra, for the most part, although there is some cocoa, rice, palm oil, and cattle output.

The Solomon Islands is a classic example of an underdeveloped country. It has limited resources, finances, transportation, and technological expertise, but a rapidly growing and largely uneducated population. Much of its budget is underwritten by grants and loans from Western countries and international agencies.

The largest and most significant island is Guadalcanal, which has the capital of Honiara (population, about 25,000) on its north coast. The Solomon Islands were a British protectorate until granted independence in 1978.

Vanuatu consists of an incomplete double chain of about 80 islands that extends across 450 miles (720 km) of ocean in a northwest-southeast direction. Although many are small, most are of the high island type, and several contain active volcanoes. Vegetation is generally luxuriant, and the climate is typical of the wet tropics, except that the southernmost islands sometimes experience cool weather. As in most of Melanesia, one or more tropical cyclones can be expected every year.

The indigenous people, now known as "ni-Vanuatu," number about 140,000. They live in villages that are widely scattered in the archipelago, mostly on the coasts. No island contains more than 20,000 people, but half a dozen have more than 5,000 each. In addition to the Melanesians, there are a few hundred Europeans, Vietnamese, and Chinese.

Until attaining independence in 1980, Vanuatu was known as New Hebrides, and had been governed as a condominium by Great Britain and France for nearly a century. English and French share with Bislama (the Vanuatu pidgin) the classification of "official" languages.

Most of the people are engaged in typical Melanesian subsistence-style garden farming, raising a variety of crops (always including coconuts), and keeping pigs for customary celebrations. There is a variety of commercial agriculture, with the value of copra production roughly equalling the combined value of all other crops. Commercial fishing and beef cattle raising are also significant.

The economy is prominently supported by foreign grants and loans, particularly from Britain, France, and Australia. In addition, Vanuatu has set out to become an international investment center and tax haven. It imposes few business taxes, has no currency or exchange controls, and companies are not required to have their records open for public scrutiny, so complete secrecy can be maintained. As a result, more than 1,000 international companies maintain registered offices in Vanuatu, and many firms that offer legal, accounting, and financial services have been attracted.

Efate is the principal, though not the largest, island. It contains the capital (Port-Vila), the two best ports, and the longest road system.

New Caledonia, third largest island in the Pacific Basin, is different both physically and culturally from the other high islands of Melanesia. The 250-mile (400-km) long mountainous backbone of this cigar-shaped island provides a rain shadow effect on the southwestern side, resulting in large areas of scrubby woodland and open savanna, rather than the luxuriant forest that dominates the landscape on the northeastern side.

The many valuable mineral deposits (primarily nickel, but also iron, chrome, manganese, cobalt, and other ores) have provided an economic stimulus that has attracted large numbers of French settlers, so that the indigenous population is no longer dominant. "Kanaks" (as the natives refer to themselves) and Europeans (virtually all French) each comprise about 40% of the population mix, with the remainder being divided among Polynesians, Indonesians, and Vietnamese.

In the past, the kanaks have lived mostly on reserves or in housing areas set aside for plantation and mine workers. Most were engaged in subsistence farming with the traditional crops of yams, taro, maize, rice, and manioc.

During the 1980s and 1990s New Caledonia has been riven with political strife that sometimes blossomed into active violence. Most of the kanaks are pressing for full independence, while most of the Europeans want to maintain the *status quo* as an overseas territory of France. Compromise has become increasingly difficult as both sides have hardened their positions.

This volatile situation has had a very depressing effect on the economy, particularly by its devastating impact on the once-flourishing tourist industry.

The focal point of New Caledonia is the administrative center of Noumea, with its large (85,000) and sophisticated urban population, huge nickel smelter, gambling casino, and beach resorts.

Although located on the periphery of the Melanesian culture zone, *Fiji* is generally considered to be a Melanesian island group. The indigenous Fijians are classed racially as Melanesians, but their culture is more strongly Polynesian than Melanesian.

Until recently, the largest component of the population has consisted of people of Indian extraction. They are descendants of indentured sugar cane workers brought from India between 1879 and 1916. For several decades, until the late 1980s, Fiji contained more ethnic Indians than it did native Fijians. These two cohorts lead very different lives, with separate languages, educational systems, electoral rolls, and (in some cases) legal restrictions. Moreover, their patterns of living, economic interests, and lifestyles are quite divergent (see Photo 10-6).

Photo 10-6 *Rice is a staple in the diet of Fiji's Indian population. Here an Indian farmer is plowing his paddy in preparation for planting rice. The scene is near Navua on Viti Levu. (TLM photo.)*

Two separate and bloodless military coups in 1987 achieved their goal of ensuring political dominance for ethnic Fijians, as well as their unstated intent to retain the elite political and economic status of the Fijian chiefly class. Since that time, there has been an exodus of several thousand Indians from Fiji, with the result that ethnic Fijians now comprise a majority of the population for the first time since 1946.

There are some 800 islands in the group, but 86% of the land and 92% of the people are encompassed in the two principal islands of Viti Levu and Vanua Levu. There is a marked contrast between the windward (southeastern) and leeward (northwestern) sides of the main islands. The former receive abundant rainfall, support dense forests, and are the favored areas for Fijian settlement, with their village-oriented subsistence agriculture. The latter get much less rainfall, are mostly unforested, and are inhabited largely by Indian sugar cane farmers.

Sugar has been the principal export and money-spinner in Fiji for more than a century, although, in recent years, the profit margin has been small. Other significant commercial products include gold, fish, copra, ginger, timber, and woodchips. Fiji faces many economic and social problems, but its economy is increasingly bolstered by expanding tourism.

Urban areas have been growing rapidly in recent years, and some 40% of the population is now urban. Suva is the capital and chief port, with a population exceeding 175,000, but two main urban corridors have emerged: Suva-Nausori in eastern Viti Levu and Nadi-Lautoka in western Viti Levu.

Polynesia. The most far-flung of the Pacific culture areas is *Polynesia*, which means "many islands." It encompasses more than a dozen principal island groups of the central and southeastern Pacific in an area shaped like a rough triangle with sides about 5,000 miles (8,000 km) long extending from Hawaii in the north to New Zealand in the south and Easter Island in the east (see Figure 10-3). Polynesia includes only 1% of the total land area of the Pacific islands, but encompasses more than 13% of the total population (excluding Hawaii). Polynesians are relatively tall, black-haired, attractive people with the lightest skin coloring of all Pacific islanders. Most cultural traits are alike throughout Polynesia. Indeed, Polynesian languages are so similar that they are often referred to as dialects, and there is much mutual intelligibility.

The kingdom of *Tonga* is a constitutional monarchy that is distinctive in the Pacific in that it was never a colonial possession, although it was under the protection of Britain for the first half of the 20th century. It includes about 150 small islands of mixed types that are arranged in three principal groups and spread over 250 miles (400 km) of ocean. The unique social structure of the country has a royalty-nobility-peasantry orientation, combined with the very strong influence of several churches, particularly Methodist. All land is owned by the crown or by certain noble families, but each adult male is given an allotment to use, provided he follows certain planning regulations. The people live in agricultural or fishing villages and are mostly engaged in subsistence activities, although copra, bananas, vanilla, squash,

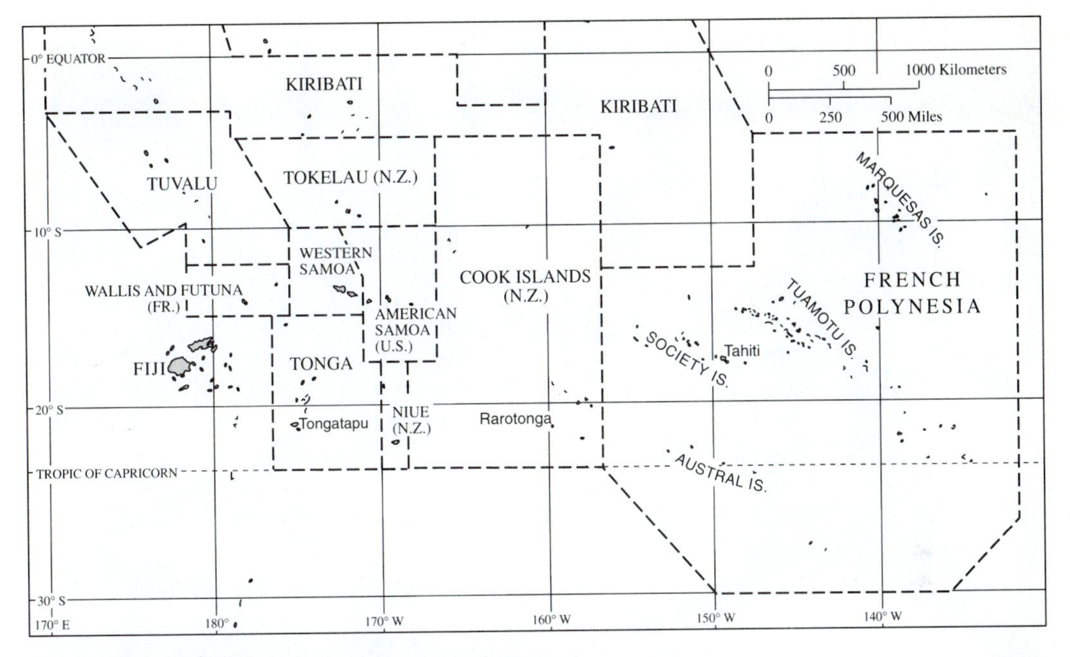

Figure 10-3 Polynesia

and watermelons are exported. The economy is mildly flourishing, and there is no public debt; Tonga can be considered a prosperous country by Pacific standards. Nearly two-thirds of the 110,000 Tongans live on the principal island of Tongatapu (see Photo 10-7).

Western Samoa achieved independence in 1962, the first Pacific island country to do so. Consisting of two medium-sized and seven small islands, the group has a population of about 175,000, almost all Samoans. The people live in small coastal villages, which are often located at stream mouths. Apia is the only town of any size. Land holdings are typically communal, and a variety of tropical subsistence crops is raised. The economy is primarily dependent upon three export crops (copra, taro, and cacao) although tourism is an increasingly important contributor. There is a steady flow of Samoan emigrants to New Zealand and the United States, from which they send money back to their relatives on the home islands. These remittances are of utmost importance to the economy of both Samoas, as they are to other Polynesian countries (particularly Tonga, the Cook Islands, Tokelau, and Niue) whose nationals have migrated in considerable numbers to New Zealand, Australia, or the United States.

American Samoa has less than one-tenth the land area and one-sixth the population of Western Samoa. Most of the people live on the island of Tutuila, although there are half a dozen other small isles in the group. The population total has remained relatively static for years, as there is a continuing migration stream to Hawaii and California (American Samoans are considered as United States nationals). Pago Pago (pronounced "Pango Pango") is

Photo 10-7 *The making of tapa is a notable handicraft and art form throughout Polynesia. Tapa is a bark cloth made from the paper mulberry tree. The painting of designs on tapa is often an important social occasion for Polynesian women. This scene is in the village of Houma on the island of Tongatapu in the Kingdom of Tonga. (TLM photo.)*

the administrative center and has perhaps the best harbor in the Pacific Basin. The economy is supported almost entirely by the catching and canning of fish (largely tuna), federal expenditures (nearly half of the work force is employed by Uncle Sam), and tourism.

The *Cook Islands* is an internally self-governing state in "free association" with New Zealand. Located southeast of Samoa, there are 15 small, scattered islands in two general groupings. The southern group is mostly mountainous and relatively fertile; the northern group consists of atolls. More than half of the 20,000 population resides on Rarotonga, the largest and most important island. The New Zealand connection is very strong: most exports (largely clothing, citrus, and bananas) are sold there, and many more Cook Islanders now live in New Zealand than in the Cooks themselves.

French Polynesia, an overseas territory of France, includes five archipelagos of some 130 islands scattered over 22° of latitude and 25° of longitude in the Southern Hemisphere. The larger islands are high and volcanic, but the greater number are small atolls (see Photo 10-8). The principal island of

Photo 10-8 *French Polynesia's Bora-Bora, a volcanic island with a surrounding reef, is one of the most beautiful of the Pacific islands. (TLM photo.)*

Tahiti encompasses about one-fourth of the total land area of the territory and provides homes for about two-thirds of the total population. Papeete is the principal urban place and is a bustling, traffic-jammed town that is the focal point of administrative, business, and tourist activities. The people of the other islands are primarily subsistence gardeners and fishermen. The only exports of significance are coconut oil, vanilla, and cultured pearls. The economy is undergirded, however, by rapidly expanding tourism (which is mostly focused upon Tahiti, Moorea, and Bora-Bora) and by governmental expenditures (particularly for nuclear testing and experimentation). The political situation of French Polynesia is in continual ferment, with a strong independence movement counterbalanced by an even more vigorous colonial sentiment.

The other island groups in Polynesia are small, unpopulous (although, in some cases, with high population densities), lacking in resources, and underdeveloped. They include:

—*Wallis and Futuna*, an overseas territory of France, consisting of about 14,000 people on two islands located between Fiji and Samoa.

—*Tuvalu* (previously called the Ellice Islands), where 9,000 people are crowded onto nine tiny atolls; the country became independent in 1978.

—*Niue* is a self-governing country in "free association" with New Zealand. It consists of a single uplifted coral island of about 4,000 population.

—*Tokelau* is a New Zealand territory located north of Samoa. It consists of three small atolls with a population of about 2,000.

—*Easter Island*, a territory of Chile, is one of the most remote inhabited places in the world. It has a population of about 2,000, of whom about 60% are Polynesians and the rest are temporary residents from Chile.

—*Pitcairn Islands* is a British dependency with a population of about 50, most of whom are descendants of the "Bounty" mutineers.

Micronesia. The 2,500 islands of Micronesia are scattered over 3,000,000 square miles (7,700,000 km^2) of ocean, almost entirely north of the Equator (see Figure 10-4). There are some high islands, but most are atolls; they range in size from small to tiny. Micronesia encompasses only about 0.3% of the total land area of the Pacific islands and only about 5% of Pacific island population. Most islands are relatively sparsely settled, but a few are densely crowded. The physical and cultural characteristics of the people are variable. There are nine regional languages, many with a variety of dialects.

The *Republic of Nauru* is the smallest (nine square miles or 24 km^2) and one of the least populous (9,000 people) countries in the world, and one of the richest per capita. It consists of a single raised coral island, the interior of which is composed mostly of the world's richest phosphate rocks. Phos-

Figure 10-4 Micronesia

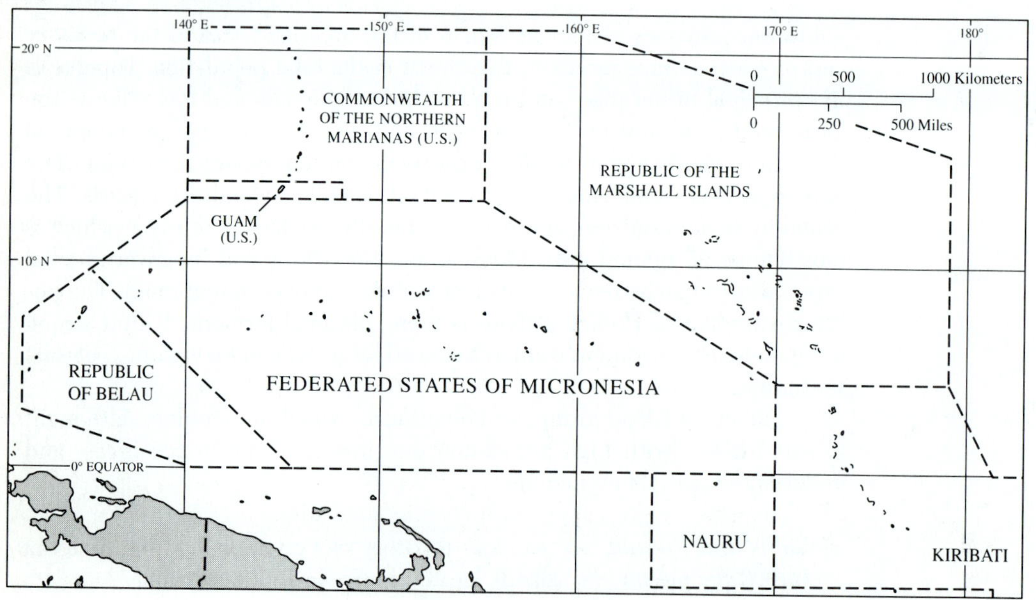

phate quarrying provides virtually the entire national income, providing the people with almost a complete welfare state in a tax-free society. About 40% of the population consists of wage-earners from other islands who have come to Nauru to work in the phosphate industry. The phosphate will be exhausted in the late 1990s, by which time it is hoped that successful external investments will provide a permanent patrimony for the people.

At the other end of the economic scale is the *Republic of Kiribati* (previously known as the Gilbert Islands), which sprawls easterly of Nauru over an area of 5,000,000 square miles (13,000,000 km^2). Its 65,000 people occupy 33 small and impoverished islands. The only significant source of income is copra and fish, although its 200-mile economic zone is so vast that considerable income is obtained from the licensing of foreign fishing vessels in Kiribati waters.

Guam is the largest island and contains about half of the total population of Micronesia. It has been a territory of the United States since it was obtained from Spain in 1898. Nearly one-third of the island's area is used by the military, and more than one-third of the island's population is from the United States mainland, principally military personnel and dependents. Nearly half of the population is derived from the island's natives, called Chamorros. In recent years, there has been a significant flow of immigrants from the Philippines. Guam has become a major air transportation hub in the western Pacific, and its tourist industry is booming; 90% of the tourists are from Japan.

Most of Micronesia (some 2,000 islands in the Caroline, Marshall, and Mariana archipelagos) was administered by the United States until the mid-1980s as a single political unit, the Trust Territory of the Pacific Islands. It was then separated into four constitutional governments, all of which retain a "relationship" with the United States. All four share the problems of vast distances, a great lack of resources, a sharp conflict of traditional and modern value systems, a lack of economic infrastructure, and political confusion. Their economies are supported almost entirely by the United States government.

The *Federated States of Micronesia* (which comprise the associated states of Kosrae, Ponape, Truk, and Yap) include several dozen atolls and a few high islands that used to be known as the Caroline Islands. The population is about 75,000.

The *Republic of Belau* (Palau) includes about 200 (mostly tiny) islands of the western Carolines, with a population of about 15,000.

The *Republic of the Marshall Islands* encompasses nearly 1,000 tiny atolls and reefs with a population of about 35,000.

The *Commonwealth of the Northern Mariana Islands* has chosen to retain closer ties with the United States (analogous to Puerto Rico). It includes a chain of 17 islands running north from Guam, with a population of about 20,000. Included are two prominent battlefield islands—Saipan and Tinian—from World War II. Tourism, primarily from Japan, is the principal economic activity.

Between the Northern Marianas and Japan are the *Ogasawara Islands* (also known as the Bonin Islands), a Japanese possession that is actually a municipal unit of Tokyo. The 30 islands have a population of about 2,000.

Wake Island is a tiny atoll north of the Marshall Islands, which functions as a major weather station and contingency air base for the United States. Its population consists of a dozen government employees and about one hundred guest workers from Thailand.

Island Issues

There is much variety in the Pacific island countries, particularly with regard to areal extent, environmental conditions, population numbers, and natural resources. This diversity shows up primarily in the larger islands, all of which are Melanesian. Papua New Guinea, for example, which is by far the largest country in the island realm, faces complex difficulties related to cultural, regional, social, and economic differences that are inconceivable on the small islands. Rural people swarm to the cities, particularly Port Moresby and Lae, seeking the good life, only to find few jobs, inadequate housing, and high prices. As a result, social dislocation is rampant, crime rates burgeon, and tribal and regional antagonisms flourish. In comparison with almost all other Pacific island countries, Papua New Guinea has both a remarkable range of natural resources and a stifling array of economic/social/political problems.

The commonalities of Pacific island life, however, are much more numerous than the differences. Everywhere the pleasures of living in a permissive environment are balanced against the difficulties of crowded conditions, limited resources, and inadequate sanitation. The economies of most countries are very circumscribed, and their dependence on external financial help (e.g., tourism, foreign aid, remittances) is very significant.

Although there are many areas of disagreement among peoples and governments of such a far-flung region, there are certain broad concerns on which agreement is widespread:

1. Oceanic resources should be wisely used, and non-Pacific island nations should pay for their exploitation of them. Each island nation claims a 200-mile (320-km) EEZ (Exclusive Economic Zone), and requires that foreign fishing vessels lease rights to fish within that zone. Enforcement is difficult because most of the countries have such a limited budget for surveillance and apprehension. However, after several confrontations and considerable controversy, the foreign nations affected (primarily Japan, both Chinas, Korea, Russia, and the United States) have recognized and adjusted to these regulations. Over-fishing is still a matter of concern, particularly with regard to enormous, indiscriminate drift nets.
2. Most Pacific island countries are dead set against the presence of nuclear material within their region, and have lined up solidly for the maintenance of a nuclear-free zone. In direct opposition to this sentiment, France has continue to explode nuclear weapons at its long-established test facility

in the Tuamotu Islands, and the USA is adamant not to confirm or deny if its naval vessels carry nuclear warheads, but the abrupt ending of the cold war between East and West eased the pressure on all parties, and makes this an issue of diminishing significance.

3. Concerns over the "Greenhouse Effect" are more notable in the Pacific islands than perhaps anywhere else in the world. Although the realities of increasing global warming are still quite debatable, these small island nations are aware of the potential. If global warming is a reality and the polar ice caps begin to melt, literally thousands of low islands would be flooded.

The Pacific islands represent only a tiny fraction of the world's land area and population, but their international interests and concerns are large indeed.

A Selected Bibliography of Basic Works

Periodicals

Australian Geographer, Geographical Society of New South Wales, Sydney.

Australian Geographic, Australian Geographic Society.

Australian Geographical Studies, Institute of Australian Geographers.

Islands Business Pacific, Suva, Fiji.

New Zealand Geographer, New Zealand Geographical Society, Christchurch.

Pacific Islands Monthly, Suva, Fiji.

Pacific Viewpoint, Victoria University of Wellington, Wellington.

Atlases

Atlas of Australian Resources. Canberra: Division of National Mapping, various dates.

Atlas of New South Wales, R.J. Harriman and E.S. Clifford (eds.). Bathurst: Central Mapping Authority of New South Wales, 1987.

Atlas of New Zealand Geography, G.J.R. Linge and R.M. Frazer. Wellington: A.H. & A.W. Reed, 1965.

Atlas of South Australia, T. Griffin and M. McCaskill (eds.). Adelaide: South Australian Government Printing Division, 1986.

Atlas of Tasmania, J.L. Davies (ed.). Hobart: Lands and Surveys Department, 1965.

Atlas of Victoria, J.S. Duncan (ed.). Melbourne: Government of Victoria, 1982.

Australians: A Historical Atlas, J.C.R. Camm and J. McQuilton (eds.). Sydney: Fairfax, Syme and Associates, 1987.

Descriptive Atlas of the Pacific Islands. T.F. Kennedy. Wellington: A.H. and A.W. Reed, 1966.

New Zealand Atlas, Ian Wards (ed.). Wellington: Government Printer, 1976.

Papua New Guinea Atlas: A Nation in Transition, David King and Stephen Ranck. Port Moresby: Robert Brown & Associates and University of Papua New Guinea, n.d. (ca. 1980).

Queensland Resources Atlas. Brisbane: Queensland Public Relations Bureau, 1976.

The AUSMAP Atlas of Australia, Ken Johnson. Cambridge: Cambridge University Press, 1992

Yearbooks and Handbooks

Handbook of Fiji. Sydney: Pacific Publications Pty. Ltd., irregular.

Handbook of Papua New Guinea. Sydney: Pacific Publications Pty. Ltd., irregular.

New Zealand Official Yearbook. Wellington: Government Printer, annual.

Pacific Islands Year Book and Who's Who. Sydney: Pacific Publications Pty. Ltd., issued about every three years.

Year Book of the Commonwealth of Australia. Canberra: Commonwealth Bureau of Census and Statistics, annual.

Books

Beadle, N.C.W., *The Vegetation of Australia*. Cambridge: Cambridge University Press, 1981.

Berndt, R.M. and C.H. Berndt (eds.), *Aborigines of the West: Their Past and Their Present*. Perth: University of Western Australia Press, 1980.

———, *The World of the First Australians*. Sydney: Ure Smith, 1977.

Birrell, E., D. Hill, and J. Stanley (eds.), *Quarry Australia: Social and Environmental Perspectives on Managing the Nation's Resources*. Melbourne: Oxford University Press, 1982.

Blainey, Geoffrey, *The Rush That Never Ended: A History of Australian Mining*. Melbourne: Melbourne University Press, 1964.

———, *Triumph of the Nomads: A History of Ancient Australia*. Melbourne: Macmillan Company of Australia, 1975.

Brookfield, H.C. and D. Hart, *Melanesia: A Geographical Interpretation of an Island World*. London: Methuen & Co., 1971.

Bunge, Frederica M. and Melinda W. Cooke (eds.), *Oceania: A Regional Study*. Washington: Government Printing Office, 2nd ed., 1985.

Burnley, I.H. *Population, Society and Environment in Australia*. Melbourne: Shillington, 1982.

Courtenay, P.P., *Northern Australia: Patterns and Problems of Tropical Development in an Advanced Country*. London: Longman, 1983.

Crocombe, R.G., *et al.*, *Land Tenure in the Pacific*. Suva: Institute of Pacific Studies, University of the South Pacific, 1980.

Day, Lincoln H. and D.T. Rowland, *How Many More Australians?: The Resource and Environmental Conflicts*. Melbourne: Longman Cheshire, 1988

Day, M.F., *Australia's Forests*. Canberra: Australian Academy of Science, 1981.

Dury, G.H. and M.I. Logan (eds.), *Studies in Australian Geography*. Melbourne: Heinemann Educational Australia, 1968.

Finlayson, H.H., *The Red Centre*. Sydney: Angus & Robertson, 1952.

Franklin, Harvey, *Cul de Sac: The Question of New Zealand's Future*. Sydney: George Allen and Unwin, 1985.

Gentilli, J., *Australian Climate Patterns*. Melbourne: Thomas Nelson Press Ltd., 1982.

——— (ed.), *Western Landscapes*. Perth: University of Western Australia Press, 1979.

Gill, A.M. (ed.), *Fire and the Australian Biota*. Canberra: Australian Academy of Science, 1981.

Goodman, Raymond, Charles Lepoani and David Morawetz, *The Economy of Papua New Guinea: An Independent Review*. Canberra: Development Studies Center, Australian National University, 1985.

Grattan, C.H., *The Southwest Pacific Since 1900*. Ann Arbor: University of Michigan Press, 1963.

———, *The Southwest Pacific to 1900*. Ann Arbor: University of Michigan Press, 1963.

Heathcote, R.L., *Australia*. London: Longman, rev. ed., 1993.

———, *Back of Bourke: A Study of Land Appraisal and Settlement in Semi-arid Australia*. Melbourne: Melbourne University Press, 1965.

Holland, P.G. and W.B. Johnston (eds.), *Southern Approaches; Geography in New Zealand*. Christchurch: New Zealand Geographical Society, 1987.

Holmes, J.H. (ed.), *Queensland: A Geographical Interpretation*. Brisbane: Royal Geographical Society of Australasia, Queensland Branch, 1988.

Jeans, D.N. (ed.), *Australia: A Geography, Volume 1: The Natural Environment*. Sydney: Sydney University Press, 2nd ed., 1986.

——— (ed.), *Australia: A Geography, Volume 2: Space and Society*. Sydney: Sydney University Press, 2nd ed., 1987.

Jennings, J.N. and J.A. Mabbutt (eds.), *Landform Studies from Australia and New Guinea*. New York: Cambridge University Press, 1967.

Lavery, H.J. (ed.), *The Kangaroo Keepers*. St. Lucia: University of Queensland Press, 1985.

Learmonth, Nancy and Andrew, *Regional Landscapes of Australia: Form, Function, Change*. Sydney: Angus Press, 1971.

Linacre, E. and J. Hobbs, *The Australian Climatic Environment.* New York: John Wiley & Sons, 1977.

Linge, G.J.R., *Canberra: Site and City.* Canberra: Australian National University Press, 1975.

———, *Industrial Awakening: A Geography of Australian Manufacturing 1788–1890.* Canberra: Australian National University Press, 1979.

Mercer, D.C., *"A Question of Balance": Natural Resource Conflict Issues in Australia.* Sydney: Federation Press, 1991.

Morton, Harry and Carol Morton Johnson, *The Farthest Corner: New Zealand—A Twice Discovered Land.* Honolulu: University of Hawaii Press, 1988.

Oliver, Douglas L., *Native Cultures of the Pacific Islands.* Honolulu: University of Hawaii Press, 1989.

———, *The Pacific Islands.* Honolulu: University of Hawaii Press, rev. ed., 1983.

Parkes, Don (ed.), *Northern Australia: The Arenas of Life and Ecosystems on Half a Continent.* Sydney: Academic Press, 1984.

Pigram, John, *Issues in the Management of Australia's Water Resources.* Melbourne: Longman Cheshire, 1986.

Powell, J.M., *An Historical Geography of Modern Australia: The Restive Fringe.* New York: Cambridge University Press, 1991.

——— and M. Williams (eds.), *Australian Space, Australian Time.* New York: Oxford University Press, 1975.

Pyne, Stephen J. *Burning Bush: A Fire History of Australia.* New York: Henry Holt and Company, 1991.

Rich, D.C., *The Industrial Geography of Australia.* London: Croom Helm, 1987.

Serventy, Vincent, *Landforms of Australia.* Sydney: Angus & Robertson, 1968.

Sharp, A., *Ancient Voyagers in Polynesia.* Sydney: Angus & Robertson, 1964.

Slayter, R.O. and R.A. Perry (eds.), *Arid Lands of Australia.* Canberra: Australian National University Press, 1969.

Smith, Jeremy (ed.), *The Unique Continent: An Introductory Reader in Australian Environmental Studies.* St. Lucia, Qld.: University of Queensland Press, 1992.

Smith, L.M., *The Aboriginal Population of Australia.* Canberra: Australian National University Press, 1980.

Solomon, R.J., *The Richest Lode: Broken Hill, 1883–1988.* Sydney: Hale and Iremonger, 1988.

Spate, O.H.K., *Australia.* New York: Frederick A. Praeger, 1962.

———, *The Pacific Since Magellan, Vol I. The Spanish Lake.* Canberra: Australian National University Press, 1981.

———, *The Pacific Since Magellan, Vol. II. Monopolists and Freebooters.* Canberra: Australian National University Press, 1983.

———, *The Pacific Since Magellan, Vol. III. Paradise Found and Lost.* Canberra: Australian National University Press, 1988.

Stevens, Graeme, *Lands in Collision: Discovering New Zealand's Past Geography.* Wellington: DSIR Science Information Publishing Centre, 1985.

Walmsley, D.J. and A.D. Sorensen, *Contemporary Australia: Explorations in Economy, Society and Geography.* Melbourne: Longman Cheshire, 2nd ed., 1993.

Ward, Gerald R. and A. Proctor (eds.), *South Pacific Agriculture: Choices and Constraints.* Canberra: Australian National University Press, 1980.

Warner, R.F. (ed.), *Fluvial Geomorphology of Australia.* Sydney: Harcourt Brace Jovanovich, 1988.

Wiens, J.J. *Atoll Environment and Ecology.* New Haven: Yale University Press, 1962.

Williams, D.B., *Agriculture in the Australian Economy.* Sydney: Sydney University Press, 1982.